A STUDENT'S POCKET COMPANION

J. RICHARD CHRISTMAN
U.S. Coast Guard Academy

to accompany

FUNDAMENTALS OF
PHYSICS

FIFTH EDITION

DAVID HALLIDAY
University of Pittsburgh

ROBERT RESNICK
Rensselaer Polytechnic Institute

JEARL WALKER
Cleveland State University

JOHN WILEY & SONS, INC.
New York   Chichester   Brisbane   Toronto   Singapore

PREFACE

A Student's Pocket Companion to *Fundamentals of Physics*, fifth edition, by Halliday, Resnick, and Walker lists the important ideas covered in each section of the text, with a few sentences about each, and gives the basic equations. It serves three purposes. First, it can be taken to class as a substitute for the text. You might want to check off the topics covered and make short notes to remind yourself of important points. A wide left margin and a notes section at the end of each chapter are provided for this purpose. Second, it can be used as a handy reference for ideas and equations while working problem assignments. Third, it can be used to review text material before an exam or when you need to recall an idea from a previous chapter.

Reading *A Student's Pocket Companion* is NOT a substitute for reading the text. Some derivations and applications are outlined in *A Pocket Student's Companion* but they are necessarily shortened. The text contains much more detail and much fuller explanations. Study the text well, preferably before class, then use *A Student's Pocket Companion* to remind yourself of the material you have studied. If it fails to jog your memory, restudy the appropriate portion of the text. A short vocabulary list is provided at the beginning of each chapter. In order to understand the material of the chapter you should know the meanings of these words and phrases. Some definitions are given in *A Student's Pocket Companion*; for other definitions you should refer to the text.

Full understanding of the ideas outlined in *A Student's Pocket Companion* will help you greatly in solving problems. However, problem solving techniques are not explicitly covered. For help in solving problems refer to the Sample Problems of the text, the full size *Student's Companion*, and the *Solutions Manual*. Also read the Problem Solving Tactics sections of the text.

Acknowledgements   Many good people at John Wiley & Sons helped put together *A Student's Pocket Companion*. Among them, Cliff Mills, Joan Kalkut, Erica Liu, and Rita Kerrigan were instrumental in the conception, design, and production. Catherine Donovan, in her usual efficient manner, carried out a myriad of essential tasks. Stuart Johnson, the current Physics Editor, has supported the latest edition. Monica Stipanov and Jennifer Bruer have each contributed in a great many ways. I am grateful to them all. I am also grateful to Karen Christman, who carefully proofread the manuscript. Very special thanks goes to Mary Ellen Christman, whose support and encouragement seem to know no bound.

J. Richard Christman
U.S. Coast Guard Academy
New London, CT   06320

TABLE OF CONTENTS

Chapter 1
MEASUREMENT

Physics is an experimental science and relies strongly on accurate measurements of physical quantities. All measurements are comparisons, either direct or indirect, with standards. This means that for every quantity you must not only have a qualitative understanding of what the quantity represents but also an understanding of how it is measured. A length measurement is a familiar example. You should know that the length of an object represents its extent in space and also that length might be measured by comparison with a meter stick, say, whose length is accurately known in terms of the SI standard for the meter. Make a point of understanding both aspects of each new quantity as it is introduced.

Important Concepts

☐ unit

☐ standard

☐ base quantity
   (base unit, base standard)

☐ International System of Units

☐ conversion factor

☐ meter

☐ second

☐ kilogram

☐ atomic mass unit

1–1 Measuring Things

☐ A **unit** is a well-defined quantity with which other quantities are compared in a measurement. Examples: the unit of length is the meter, the unit of time is the second, the unit of mass is the kilogram.

☐ Some units are defined in terms of others. For example, the unit for speed is the meter per second. Others are **base units** and are defined in terms of **standards**. Ideally, a standard should be accessible and invariable.

☐ A system of units consists of a unit for each physical quantity, organized so that all can be derived from a small number of independent base units.

1–2 The International System of Units

- [ ] This system is called the SI system (previously, the metric system).

- [ ] The three International System <u>base</u> <u>units</u> used in mechanics are:

length:	meter (abbreviation: m)
time:	second (abbreviation: s)
mass:	kilogram (abbreviation: kg)

- [ ] SI prefixes are used to represent powers of ten. The following are used the most:

Prefix	Power of Ten	Symbol
kilo:	10^3	k
mega:	10^6	M
centi:	10^{-2}	c
milli:	10^{-3}	m
micro:	10^{-6}	μ
nano:	10^{-9}	n
pico:	10^{-12}	p

Memorize them. When evaluating an algebraic expression, substitute the value using the appropriate power of ten. That is, for example, if a length is given as $25\,\mu$m, substitute 25×10^{-6} m. One catch: the SI unit for mass is the <u>kilogram</u>. Thus, a mass of 25 kg is substituted directly, while a mass of 25 g is substituted as 25×10^{-3} kg.

1–3 Changing Units

- [ ] There are usually several common units for each physical quantity. For example, length can be measured in meters, feet, yards, miles, light years, and other units.

☐ A quantity given in one unit is converted to another by multiplying by a **conversion factor**. Some conversion factors are listed in Appendix D of the text.

☐ Carefully study Section 1–3 to see how a quantity given in one unit is converted to another. Cultivate the good habit of saying the words associated with a conversion. Suppose you want to convert 50 ft to meters. Appendix D tells you that 1 ft is equivalent to 0.3048 m. Say "Since 1 ft is equivalent to 0.3048 m, then 50 ft must be equivalent to $(50 \, \text{ft}) \times (0.3048 \, \text{m/ft}) = 15 \, \text{m}$".

1–4 Length

☐ The SI standard for the meter is the distance traveled by light during a time interval of $1/299, 792, 458$ s.

☐ This makes the speed of light exactly $299, 792, 458$ m/s.

☐ Table 1–3 gives some lengths. Note the wide range of values.

1–5 Time

☐ The SI standard for the second is the time taken for exactly $9, 192, 631, 770$ vibrations of a certain light emitted by cesium-133 atoms.

☐ Table 1–4 lists some time intervals. Note the wide range of values.

1–6 Mass

☐ The SI standard for the kilogram is the mass of a platinum-iridium cylinder carefully stored at the International Bureau of Weights and Measures near Paris, France.

☐ A second mass unit is the **atomic mass unit**, abbreviated u and defined so the mass of a carbon-12 atom is exactly 12 u. $1 \, \text{u} = 1.6605402 \times 10^{-27}$ kg.

☐ Table 1–5 lists some masses. Note the wide range of values.

NOTES:

Chapter 2
MOTION ALONG A STRAIGHT LINE

This chapter introduces you to some of the concepts used to describe motion; most important are those of position, velocity, and acceleration. Pay particular attention to their definitions and to the relationships between them.

Important Concepts

- ☐ particle
- ☐ coordinate axis
- ☐ origin
- ☐ coordinate
- ☐ displacement
- ☐ average velocity
- ☐ average speed
- ☐ (instantaneous) velocity

- ☐ (instantaneous) speed
- ☐ average acceleration
- ☐ (instantaneous) acceleration
- ☐ motion with constant acceleration
- ☐ free-fall acceleration
- ☐ free-fall motion

2–1   Motion

☐   In this section of the text, objects are treated as particles. A **particle** has no extent in space and has no internal parts that can move relative to each other. It may have other properties, such as mass.

☐   An extended object can be treated as a particle if all points in it move along parallel lines. It cannot rotate and it cannot deform. If an extended object can be treated as a particle, we may pick one point on the object and follow its motion. The position of a crate, for example, means the position of the point on the crate we have chosen to follow, perhaps one of its corners.

2–2   Position and Displacement

☐   The motion of a particle in one dimension can be described by giving its **coordinate** x as a function of time t. Draw a **coordinate axis** along the line of motion of the particle and select one point on the axis to be the **origin**. The distance from the origin to the particle is the magnitude of the coordinate. The coordinate is positive if it is on the side of the origin designated positive and negative if it is on the side designated negative.

☐   You must carefully distinguish between an *instant* of time and an *interval* of time. The symbol t represents an instant and has no extension. Thus, t might be *exactly* 12 min, 2.43 s after noon on a certain day. At any other time, no matter how close, t has a different value. On the other hand, an interval extends from some initial time to some final time: *two* instants of time are required to describe it. Note that a value of the time may be positive or negative, depending on whether the instant is after or before the instant designated as $t = 0$.

☐   Similarly, a value of the coordinate x specifies a *point* on the x axis. It has no extension in space.

☐   A **displacement** is a difference in two coordinates. If a particle goes from x_1 to x_2 during some interval of time, its displacement during that interval is $\Delta x = x_2 - x_1$. Notice that the *initial* coordinate is subtracted from the *final* coordinate. This definition is valid no matter what the signs of x_1 and x_2.

☐   Two particles that start at x_1 and end at x_2 have the same displacement no matter what their motions. The magnitude of the displacement during a time interval may be different from the distance traveled during the interval. The difference is pronounced if, for example, the particle moves back and forth several times in the interval.

2–3 Average Velocity and Average Speed

□ If a particle goes from x_1 at time t_1 to x_2 at time t_2, its **average velocity** $\overline{v}$ in the interval from t_1 to t_2 is given by

$$\overline{v} = \frac{x_2 - x_1}{t_2 - t_1} = \frac{\Delta x}{\Delta t},$$

where $\Delta x = x_2 - x_1$ and $\Delta t = t_2 - t - 1$.

□ If you are given the function $x(t)$ and are asked for the average velocity in some interval from t_1 to t_2, first evaluate the function for $t = t_1$ to find x_1, then evaluate the function for $t = t_2$ to find x_2 and finally substitute the values into the defining equation.

□ On a graph of x vs. t the average velocity over the interval from t_1 to t_2 is given by the slope of the line from t_1, x_1 to t_2, x_2. On the graph below $t_1 = 2.0\,\text{s}$ and $t_2 = 8.0\,\text{s}$. The average velocity in this interval is the slope of the dotted line.

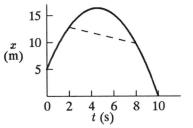

□ A downward sloping line (from left to right) has a negative slope and indicates a negative average velocity. An upward sloping line has a positive slope and indicates a positive velocity.

□ Carefully distinguish between average velocity and **average speed**. The average speed over a time interval Δt is defined by

$$\overline{s} = \frac{d}{\Delta t},$$

where d is distance traveled in the interval. This may be quite different from the displacement if the particle moves back and forth during the interval.

2–4   Instantaneous Velocity and Speed

☐   The **instantaneous velocity** is the velocity at an instant of time (not over an interval). It is defined as the limiting value of the average velocity in an interval as the interval shrinks to zero. The term "velocity" means instantaneous velocity. The adjective "instantaneous" is implied.

☐   A positive velocity means that the particle is traveling in the positive x direction. Similarly, a negative velocity means that the particle is traveling in the negative x direction.

☐   If the function $x(t)$ is known, the velocity at any time t_1 is found by finding the derivative of $x(t)$ with respect to t and evaluating the result for $t = t_1$. The velocity at any time t is given by

$$v(t) = \frac{\mathrm{d}x(t)}{\mathrm{d}t} .$$

On a graph of x vs. t, the velocity at any time t_1 is the slope of the straight line that is tangent to the curve at the point corresponding to $t = t_1$. On the graph below the slope of the slanted dotted line gives the velocity at time $= 7.0\,\mathrm{s}$.

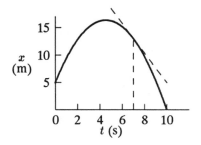

☐ The **instantaneous speed** (or just plain **speed**) of a particle is the magnitude of its velocity. If the velocity is -5.0 m/s, for example, then the speed is $+5.0$ m/s.

2–5   Acceleration

☐ If the velocity of a particle changes from v_1 at time t_1 to v_2 at a later time t_2, then its **average acceleration** $\overline{a}$ over the interval from t_1 to t_2 is given by

$$\overline{a} = \frac{v_2 - v_1}{t_2 - t_1} = \frac{\Delta v}{\Delta t} \,,$$

where $\Delta v = v_2 - v_1$ and $\Delta t = t_2 - t_1$. Notice that the velocity at the *beginning* of the interval is subtracted from the velocity at the *end*. Also notice that *instantaneous* velocities appear in the definition.

☐ If the function $x(t)$ is given, first differentiate it with respect to t to find the velocity as a function of time, then evaluate the result for times t_1 and t_2 to find values for v_1 and v_2. Substitute the values into the defining equation for the average acceleration.

☐ On a graph of v vs t, the average acceleration over the interval from t_1 to t_2 is the slope of the line from t_1, v_1 to t_2, v_2.

☐ The **instantaneous acceleration** gives the acceleration of the particle at an instant of time, rather than over an interval. It is the limiting value of the average acceleration as the interval becomes zero. "Acceleration" and "instantaneous acceleration" mean the same thing.

☐ If the function $v(t)$ is known, the instantaneous acceleration is found by differentiating it with respect to t. If the function $x(t)$ is known, the instantaneous acceleration is found by differentiating it twice with respect to t:

$$a(t) = \frac{\mathrm{d}v}{\mathrm{d}t} = \frac{\mathrm{d}^2 x}{\mathrm{d}t^2} \,.$$

☐ On a graph of v vs. t, the instantaneous acceleration at any time t_1 is the slope of the straight line that is tangent to the curve at $t = t_1$.

☐ Note that a positive acceleration does not necessarily mean the particle speed is increasing and a negative acceleration does not necessarily mean the particle speed is decreasing. The speed increases if the velocity and acceleration have the same sign and decreases if they have opposite signs, no matter what the signs.

☐ At the instant a particle is momentarily at rest its acceleration is not necessarily 0. If, at an instant, the velocity is zero but the acceleration is not, then in the next instant the velocity is not zero and the particle is moving.

2–6 Constant Acceleration: A Special Case

☐ If a particle is moving along the x axis with constant acceleration a, its coordinate and velocity are given as functions of time t by

$$x(t) = x_0 + v_0 t + \tfrac{1}{2}at^2 \qquad \text{and} \qquad v(t) = v_0 + at$$

where x_0 is its coordinate at $t = 0$ and v_0 is its velocity at $t = 0$. Notice that the second equation is the derivative of the first.

☐ Eq. 2–16 of the text is also extremely useful. It is

$$v^2 = v_0^2 + 2a(x - x_0).$$

It gives the velocity as a function of the coordinate.

☐ Problems involving motion with constant acceleration are worked by solving these equations for the unknowns. Usually two events are described in the problem statement. Select the time to be zero at one of the events. x_0 is the coordinate of the particle and v_0 is its velocity then. The other event occurs at some time t. x is the coordinate and v is the velocity then. The other quantity that enters is the acceleration a. Of the six quantities that occur in the equations, four are usually given or implied and two are unknown.

2–7 Another Look at Constant Acceleration

☐ Integration can be used to derive the equations for motion with constant acceleration. If a is the acceleration, then the velocity is given by $v(t) = \int a \, dt + C = at + C$, where C is a constant of integration. Its physical meaning is the velocity at $t = 0$. To see this, just set $t = 0$ in the equation above. Thus, $C = v_0$ and $v(t) = v_0 + at$.

☐ The coordinate is given by $x(t) = \int v(t) \, dt + C' = \int (v_0 + at) \, dt + C' = v_0 t + \frac{1}{2}at^2 + C'$, where C' is a constant of integration. It is the coordinate when $t = 0$. Thus, $C' = x_0$ and $x(t) = x_0 + v_0 t + \frac{1}{2}at^2$.

2–8 Free-Fall Acceleration

☐ Every object near the surface of the Earth, if acted on only by the gravitational force of the Earth, has the same acceleration, called the **free-fall acceleration** and denoted by g. It is downward, toward the Earth. The value of g is independent of the mass or shape of the object. It varies slightly from place to place on the Earth.

☐ In the absence of air resistance, a ball (or any other object) thrown upward has the same acceleration at all times: during its upward flight, during its downward fall, and even at the very top of its trajectory.

☐ The constant acceleration equations are revised somewhat to deal with free-fall motion. A vertical y axis is drawn with the upward direction being positive and the equations are written

$$y(t) = y_0 + v_0 t - \frac{1}{2}gt^2, \qquad v(t) = v_0 - gt,$$

and
$$v^2 = v_0^2 - 2g(y - y_0).$$

They can be obtained from the equations given previously for motion along an x axis by substituting y for x and $-g$ for a.

☐ Problems dealing with free fall are solved in exactly the same way as other problems dealing with constant acceleration.

2–9 The Particles of Physics

☐ **Quantum physics** deals with phenomena at the atomic and subatomic levels. Matter ultimately is not continuous but is made up of quanta of matter, called particles. Other quantities, such as energy, are also "lumpy" (quantized) at the atomic level and below.

☐ An **atom** consists of a central, highly compact **nucleus**, surrounded by one or more **electrons** and is held together by electrical forces.

☐ An atomic nucleus is composed of electrically neutral particles, called **neutrons**, and electrically charged particles, called **protons**. The protons repel each other electrically, but all neutrons and protons in a nucleus attract each other via a strong nuclear force. The neutrons provide the glue that holds a nucleus together.

☐ Protons, neutrons, and many other particles (but not electrons) are not fundamental but rather are composed of particles known as **quarks**.

☐ Quarks and **leptons** (of which the electron is one) seem to be the fundamental building blocks of nature.

NOTES:

Chapter 3
VECTORS

You will deal with vector quantities throughout the course. In this chapter, you will learn about their properties and how they are manipulated mathematically. A solid understanding of this material will pay handsome dividends later.

Important Concepts

☐ vector
☐ scalar
☐ component of a vector
☐ unit vector
☐ vector addition
☐ negative of a vector

☐ vector subtraction
☐ multiplication of a vector by a scalar
☐ scalar product of two vectors
☐ vector product of two vectors

3–1 Vectors and Scalars

☐ A **vector** quantity has a direction as well as a magnitude and obeys the rules of vector addition (discussed later in this chapter). In contrast, a **scalar** quantity has only a magnitude and obeys the rules of ordinary arithmetic. Displacement, velocity, acceleration, and force are vector quantities; mass, speed, charge, and temperature are scalar quantities.

☐ A vector is represented graphically by an arrow in the direction of the vector, with length proportional to the magnitude of the vector (according to some scale). As an algebraic symbol, a vector is written in boldface (**a**, for example) or with an arrow over the symbol ($\vec{a}$). The magnitude of **a** is written a, in italics (or not bold and without an arrow) or as $|\mathbf{a}|$ (or $|\vec{a}|$). Be sure you follow this convention.

It helps you distinguish vectors from scalars and components of vectors. It helps you communicate properly with your instructors and exam graders. Do *not* write $a + b$ when you mean $\vec{a} + \vec{b}$, for example. They have entirely different meanings!

☐ Displacement vectors are used as examples of vectors in this chapter. A displacement vector is a vector from the position of a particle at the beginning of a time interval to its position at the end of the interval. Note that a displacement vector tells us nothing about the path of the object, only the relationship between the initial and final positions. When you need an example to illustrate addition or subtraction of vectors, think of displacement vectors.

3–2 Adding Vectors: Graphical Method

☐ If $\mathbf{d_1}$ is the displacement vector from point A to point B and $\mathbf{d_2}$ is the displacement vector from point B to point C, then the sum (written $\mathbf{d_1} + \mathbf{d_2}$) is the displacement vector from A to C:

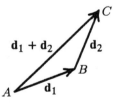

☐ To add two vectors, place the tail of the second vector at the head of the first, then draw the vector from the tail of the first to the head of the second. The order in which you draw the vectors is not important: $\mathbf{a} + \mathbf{b} = \mathbf{b} + \mathbf{a}$. It is important that the tail of one be at the head of the other and that the resultant vector be from the "free" tail to the "free" head, as in the diagram above.

☐ Except in special circumstances, the magnitude of the resultant vector is *not* the sum of the magnitudes of the vectors entering the sum and the direction of the resultant vector is *not* in the direction of any of the vectors entering the sum.

☐ Remember you can reposition a vector as long as you do not change its magnitude and direction. Thus, if two vectors you wish to add graphically do not happen to be placed with the tail of one at the head of the other, simply move one into the proper position.

☐ The idea of the negative of a vector is used to define vector subtraction. The negative of a vector is a vector that is parallel to the original vector but in the opposite direction:

☐ Vector subtraction is defined by $a - b = a + (-b)$. That is, you add a and $-b$.

☐ Notice that vector subtraction is defined so that if $a + b = 0$, then $a = -b$ and if $c = a + b$, then $a = c - b$. Just subtract b from both sides of each equation. Vector subtraction is clearly useful for solving vector equations.

3–3 Vectors and Their Components

☐ To find the x **component** of a vector, draw lines from the head and tail to the x axis, both perpendicular to the axis. The x component of the vector is the projection of the vector on the axis and is indicated by the separation of the points where the lines meet the axis. Similarly, to find the y component draw lines from the head and tail to the y axis. The diagram below shows the components of a vector.

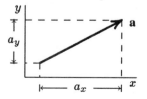

☐ The components of a vector are not vectors themselves but they can be either positive or negative. The vector in the diagram above has positive x and y components. The vector in the diagram below has a positive x component and a negative y component.

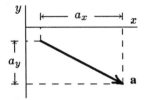

☐ Given the magnitude a of a vector and the angle θ it makes with the x axis, you can find the components using

$$a_x = a \cos \theta \qquad \text{and} \qquad a_y = a \sin \theta.$$

For these expressions to be valid, θ must be measured counterclockwise from the positive x direction:

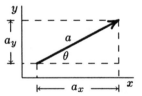

If θ is between 0 and 90°, both the x and y components are positive; if θ is between 90° and 180°, the x component is negative and the y component is positive; if θ is between 180° and 270°, both the x and y components are negative; if θ is between 270° and 360°, the x component is positive and the y component is negative.

☐ You must also be able to find the magnitude and orientation of a vector when you are given its components. Suppose a vector **a** lies in the xy plane and its components a_x

and a_y are given. In terms of the components, the magnitude of **a** is given by

$$a = \sqrt{a_x^2 + a_y^2}$$

and the angle **a** makes with the positive x direction is given by

$$\tan \theta = \frac{a_y}{a_x} .$$

The magnitude is always positive. There are two possible solutions to the equation for θ, the one given by your calculator and that angle plus 180°. You must look at the orientation of the vector to see which makes physical sense. For example, if $a_y/a_x = 0.50$, then $\theta = 26.6°$ or 206.6°. In the first case, both components are positive, while in the second, both are negative.

3–4 Unit Vectors

☐ The **unit vectors i, j**, and **k** are used when a vector is written in terms of its components. These vectors have magnitude 1 and are in the positive x, y, and z directions respectively.

☐ a_x **i** is a vector parallel to the x axis with x component a_x, a_y **j** is a vector parallel to the y axis with y component a_y, and a_z **k** is a vector parallel to the z axis with z component a_z. The vector **a** is given by

$$\mathbf{a} = a_x\,\mathbf{i} + a_y\,\mathbf{j} + a_z\,\mathbf{k} ,$$

where the rules of vector addition apply.

☐ The symbols $\hat{\imath}$, $\hat{\jmath}$, and $\hat{k}$ are used to hand write the unit vectors.

☐ Units are associated with the components a_x, a_y, and a_z of a vector but *not* with the unit vectors **i, j**, and **k**. Thus, the same unit vectors can be used to write any vector, no matter what its units.

3–5 Adding Vectors by Components

☐ Suppose **c** is the sum of two vectors **a** and **b** (i.e. **c** = **a** + **b**). In terms of the components of **a** and **b**: $c_x = a_x + b_x$, $c_y = a_y + b_y$, and $c_z = a_z + b_z$.

☐ Suppose **c** is the negative of **a** (i.e. **c** = −**a**). In terms of the components of **a**: $c_x = -a_x$, $c_y = -a_y$, and $c_z = -a_z$.

☐ Suppose **c** is the difference of two vectors **a** and **b** (i.e. **c** = **a** − **b**). In terms of the components of **a** and **b**: $c_x = a_x - b_x$, $c_y = a_y - b_y$, and $c_z = a_z - b_z$.

3–6 Vectors and the Laws of Physics

☐ Many of the laws of physics are written in vector notation. For example, Newton's second law of motion tells us that a single force **F** acting on an object of mass m produces an acceleration **a** according to **F** = m**a**. Notice that both the right and left hand sides of the equation are vectors. The advantage to writing the law in vector form is that the equation is independent of the coordinate system used. In fact, no coordinate system need be specified when the equation is written.

☐ The vector equation **a** = **b** stands for $a_x = b_x$, $a_y = b_y$, and $a_z = b_z$, no matter what the orientation of the coordinate system used to find the components. For different problems, different orientations of the coordinate system are convenient. The diagram shows two possible systems that might be used to resolve the same vector into components.

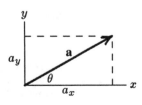 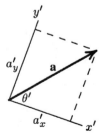

The x' and y' components are given by $a'_x = a \cos \theta'$ and $a'_y = a \sin \theta'$. The angles θ and θ' are related by $\theta = \theta' + \phi$, where ϕ is the angle by which the primed coordinate system is rotated from the unprimed. It is positive for a counterclockwise rotation. (For the coordinate systems in the diagrams above, ϕ is negative.)

3–7 Multiplying Vectors

☐ Vectors can be multiplied by scalars. Let **a** be a vector and s a scalar. Then, s**a** is a vector. If s is positive, its direction is the same as that of **a** and its magnitude is sa. If s is negative, the direction of s**a** is opposite that of **a** and its magnitude is $|s|a$.

☐ If **b** = s**a**, then in terms of components $b_x = sa_x, b_y = sa_y$, and $b_z = sa_z$.

☐ Division of a vector by a scalar is, of course, just multiplication by the reciprocal of the scalar.

☐ The **scalar product** (or dot product) of two vectors is defined in terms of the magnitudes of the two vectors and the angle between them when they are drawn with their tails at the same point: $\mathbf{a} \cdot \mathbf{b} = ab \cos \phi$. The geometry is shown in the diagram below.

☐ Remember that the scalar product of two vectors is a scalar and has no direction associated with it.

☐ $\mathbf{a} \cdot \mathbf{b}$ is positive if ϕ is between 0 and 90°; it is negative if ϕ is between 90° and 180°; it is zero if $\phi = 90°$.

☐ In terms of components, $\mathbf{a} \cdot \mathbf{b} = a_x b_x + a_y b_y + a_z b_z$.

☐ You should notice that $\mathbf{a} \cdot \mathbf{b}$ can be interpreted as the magnitude of **a** times the component of **b** along the direction of

a or as the magnitude of **b** times the magnitude of **a** along the direction of **b**. This will be a useful interpretation when you study work in Chapter 7.

☐ The **vector product** (or cross product) of two vectors, written **a**×**b**, is a vector. Its magnitude is given by

$$|\mathbf{a} \times \mathbf{b}| = ab \sin \phi,$$

where ϕ is the angle between **a** and **b** when they are drawn with their tails at the same point. ϕ is always in the range from 0 to 180° and the magnitude of the vector product is always positive.

☐ The direction of the vector product is always perpendicular to each of the factors. To find the direction of **a**×**b**, draw the vectors with their tails at the same point and pretend there is a hinge at that point. Curl the fingers of your right hand so they rotate **a** into **b**. Your thumb will then point in the direction of **a**×**b**. Note that **b**×**a** = −**a**×**b**.

☐ The vector product of two vectors with given magnitudes is zero if the vectors are parallel or antiparallel ($\phi = 0$ or 180°); it has its maximum value if they are perpendicular to each other ($\phi = 90°$).

NOTES:

Chapter 4
MOTION IN TWO AND THREE DIMENSIONS

The ideas of position, velocity, and acceleration that were introduced earlier in connection with one-dimensional motion are now extended. You should pay close attention to the definitions and relationships discussed in this chapter. Three applications are discussed: projectile motion, circular motion, and relative motion.

Important Concepts

☐ position vector   ☐ average acceleration

☐ displacement vector   ☐ (instantaneous) acceleration

☐ average velocity   ☐ projectile motion

☐ (instantaneous) velocity   ☐ uniform circular motion

☐ speed   ☐ relative motion

4–2 Position and Displacement

☐ The fundamental concept used to describe the motion of a particle moving in two or three dimensions is its **position vector**. The tail of this vector is always at the origin and at any instant the head is at the particle. The cartesian components of the position vector are the coordinates of the particle. As the particle moves its position vector changes and so is a function of time.

☐ A **displacement vector** is used to describe a change in a position vector. If the particle has position vector $\mathbf{r}_1$ at time t_1 and position vector $\mathbf{r}_2$ at a later time t_2, then the displacement vector for this interval is $\Delta \mathbf{r} = \mathbf{r}_2 - \mathbf{r}_1$. If the particle has coordinates x_1, y_1, z_1 at time t_1 and coordinates x_2, y_2, z_2 at time t_2, then the components of the displacement vector are given by $(\Delta \mathbf{r})_x = \Delta x = x_2 - x_1$,

$(\Delta \mathbf{r})_y = \Delta y = y_2 - y_1$, and $(\Delta \mathbf{r})_z = \Delta z = z_2 - z_1$. When using these equations, pay close attention to the order of the terms: a displacement vector is a position vector at a *later* time minus a position vector at an *earlier* time.

4–3 Velocity and Average Velocity

☐ In terms of the displacement vector $\Delta \mathbf{r}$, the **average velocity** of the particle in the interval from t_1 to t_2 is

$$\bar{\mathbf{v}} = \frac{\Delta \mathbf{r}}{\Delta t},$$

where $\Delta t = t_2 - t_1$. The average velocity has components given by

$$\bar{v}_x = \frac{\Delta x}{\Delta t}, \qquad \bar{v}_y = \frac{\Delta y}{\Delta t}, \qquad \text{and} \qquad \bar{v}_z = \frac{\Delta z}{\Delta t}.$$

☐ To use the definition to calculate the average velocity over the time interval from t_1 to t_2, you must know the position vector for the beginning and end of the interval. Just as for one–dimensional motion, it is important to realize that the components of the position vector are coordinates and represent points on a coordinate axis. They are not necessarily related in any way to the distance traveled by the particle.

☐ The **instantaneous velocity v** at any time t is the limit of the average velocity over a time interval that includes t, as the duration of the interval becomes vanishingly small. In terms of the position vector, it is given by the derivative

$$\mathbf{v} = \frac{d\mathbf{r}}{dt}.$$

In terms of the particle coordinates, its components are

$$v_x = \frac{dx}{dt}, \qquad v_y = \frac{dy}{dt}, \qquad \text{and} \qquad v_z = \frac{dz}{dt}.$$

The term "velocity" means the same as "instantaneous velocity".

☐ You should be aware that the instantaneous velocity, unlike the average velocity, is associated with a single instant of time. At any other instant, no matter how close, the instantaneous velocity might be different.

☐ To use the definition to calculate the instantaneous velocity, you must know the position vector as a function of time. This is identical to knowing the coordinates as functions of time. You should remember that the instantaneous velocity vector at any time is tangent to the path at the position of the particle at that time. If you are asked for the direction the particle is traveling at a certain time, you automatically calculate the components of its velocity for that time.

☐ **Speed** is the magnitude of the instantaneous velocity and, if the velocity components are given, can be calculated using

$$v = \sqrt{v_x^2 + v_y^2 + v_z^2}.$$

4–4 Acceleration and Average Acceleration

☐ In terms of the velocity $\mathbf{v}_1$ at time t_1 and the velocity $\mathbf{v}_2$ at a later time t_2, the **average acceleration** over the interval from t_1 to t_2 is given by

$$\bar{\mathbf{a}} = \frac{\mathbf{v}_2 - \mathbf{v}_1}{t_2 - t_1} = \frac{\Delta \mathbf{v}}{\Delta t}.$$

In terms of velocity components, the components of the average acceleration are

$$\bar{a}_x = \frac{\Delta v_x}{\Delta t}, \qquad \bar{a}_y = \frac{\Delta v_y}{\Delta t}, \qquad \text{and} \qquad \bar{a}_z = \frac{\Delta v_z}{\Delta t}.$$

☐ To use the definition to calculate the average acceleration over the interval from t_1 to t_2, you must know the velocity at the beginning and end of the interval. This may mean you must differentiate the position vector with respect to time.

☐ The **instantaneous acceleration a** at any time t is the limit of the average acceleration over an interval that includes t, as the duration of the interval becomes vanishingly small. In terms of the velocity vector, it is given by

$$\mathbf{a} = \frac{d\mathbf{v}}{dt}.$$

In terms of the velocity components, its components are

$$a_x = \frac{dv_x}{dt}, \qquad a_y = \frac{dv_y}{dt}, \qquad \text{and} \qquad a_z = \frac{dv_z}{dt}.$$

☐ To use the definition to calculate the acceleration, you must know the velocity vector as a function of time. The terms "instantaneous acceleration" and "acceleration" mean the same thing.

☐ A non-zero velocity indicates that the position vector of the particle is changing with time. A non-zero acceleration indicates that the velocity vector of the particle is changing with time. Remember that these changes may be changes in magnitude, in direction, or both.

4–5 Projectile Motion

☐ If air resistance is negligible, a projectile simultaneously undergoes free-fall vertical motion and constant-velocity horizontal motion. The acceleration of the projectile has magnitude g (the free-fall acceleration) and is downward, toward the Earth. Its horizontal component is zero. Note that the vertical component is constant. Once the projectile is launched its acceleration does not change until it hits the ground or a target.

☐ Given the initial position and velocity of the projectile the kinematic equations can be used to predict its position at all times (until it hits something).

☐ The initial velocity has two nonzero components. If the y axis is positive in the upward direction and the x axis is horizontal in the plane of the trajectory, then $\mathbf{v}_0 = v_{0x}\,\mathbf{i} + v_{0y}\,\mathbf{j}$.

$\square$ If v_0 is the initial speed and θ_0 is the angle between the initial velocity vector and the horizontal, then $v_{0x} = v_0 \cos \theta_0$ and $v_{0y} = v_0 \sin \theta_0$. θ_0 (and v_{0y}) are negative if the projectile is launched downward.

4–6 Projectile Motion Analyzed

$\square$ If the coordinate axes are as described above, then the coordinates and velocity components at time t are given by

$$x(t) = x_0 + (v_0 \cos \theta_0)t, \qquad v_x(t) = v_0 \cos \theta_0,$$
$$y(t) = y_0 + (v_0 \sin \theta_0)t - \tfrac{1}{2}gt^2, \quad v_y(t) = v_0 \sin \theta_0 - gt.$$

Note that the equations for x and v_x describe motion with constant velocity and that the equations for y and v_y describe free-fall motion (with constant acceleration $-g$).

$\square$ There are two special conditions you should remember. At the highest point of its trajectory the velocity of a projectile is horizontal and $v_y = 0$. To find the time when the projectile is at its highest point, solve $0 = v_0 \sin \theta_0 - gt$ for t. When a projectile returns to the original launch height, $y = y_0$. To find the time when the projectile returns to the launch height, solve $0 = (v_0 \sin \theta_0)t - \tfrac{1}{2}gt^2$ for t. The magnitude of the displacement from the launch point to the point at launch height is called the horizontal range of the projectile.

4–7 Uniform Circular Motion

$\square$ A particle in **uniform circular motion** moves around a circle with constant speed. Remember that the velocity vector is always tangent to the path and therefore continually changes direction. This means the acceleration is *not* zero.

$\square$ If the radius of the circle is r and the speed is v, the acceleration of the particle has magnitude

$$a = \frac{v^2}{r}$$

and always points toward the center of the circle. The direction of the acceleration continually changes as the particle moves around the circle.

☐ The term "centripetal acceleration" is applied to this acceleration to indicate its direction. You should not forget it is the rate of change of velocity, as are all accelerations.

4–8 Relative Motion in One Dimension

☐ The position or velocity of a particle is measured using two coordinate systems (or reference frames) that are moving relative to each other. Given the position and velocity of the particle in one frame and the relative position and velocity of the frames, you should be able to calculate the position and velocity in the other frame.

☐ Let x_{PA} be the coordinate of a particle P as measured in frame A (see the diagram below), let x_{PB} be the coordinate as measured in frame B, and let x_{BA} be the coordinate of the origin of frame B as measured in frame A. Then, $x_{PA} = x_{PB} + x_{BA}$.

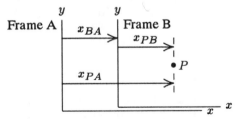

☐ Differentiate the coordinate equation with respect to time to obtain the relationship between the velocity of the particle as measured in one frame and the velocity as measured in the other. The result is $v_{PA} = v_{PB} + v_{BA}$. Here v_{PA} is the velocity as measured in A, v_{PB} is the velocity as measured in B, and v_{BA} is the velocity of frame B as measured in frame A.

☐ Take care with the subscripts. The first symbol in a subscript names an object (the particle or the origin of frame B) and the second names the coordinate frame used to measure the position or velocity of the object. You should say all the words as you read the symbols. That is, when you see x_{BA} you should say "the position vector of the origin of frame B relative to the origin of frame A." You will then get acquainted with the notation fast and won't get it mixed up later.

4–9 Relative Motion in Two Dimensions

☐ According to the diagram below, at any instant of time the position of the particle P relative to the origin of coordinate frame A is given by $\mathbf{r}_{PA} = \mathbf{r}_{PB} + \mathbf{r}_{BA}$ in terms of its position $\mathbf{r}_{PB}$ relative to the origin of frame B and the position vector $\mathbf{r}_{BA}$ of the origin of frame B relative to the origin of frame A.

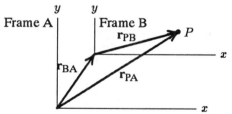

☐ The above expression can be differentiated with respect to time to obtain $\mathbf{v}_{PA} = \mathbf{v}_{PB} + \mathbf{v}_{BA}$ for the velocity of the particle in frame A in terms of the particle velocity $\mathbf{v}_{PB}$ in frame B and the velocity $\mathbf{v}_{BA}$ of frame B as measured in frame A. This expression is valid even if the two frames are accelerating relative to each other.

☐ Airplanes flying in moving air or ships sailing in moving water are often used as examples of relative motion. The airplane or ship is the particle, one coordinate frame moves with the air or water, and the other coordinate frame is fixed to the earth. The heading of the airplane or ship is in the direction of its velocity as measured relative to the air or water, *not* relative to the ground.

4–10   Relative Motion at High Speeds

☐ If an object is moving at nearly the speed of light or if we compare its velocity as measured in two reference frames that are moving at nearly the speed of light relative to each other, then the results given above fail and we must use a relativistically correct equation. Consider an object P that is moving along the x axis with velocity v_{PB}, as measured in reference frame B. If B is moving along the x axis with velocity v_{BA}, as measured in another frame A, then the velocity of P, as measured in A, is given by

$$v_{PA} = \frac{v_{PB} + v_{BA}}{1 + v_{PB}v_{BA}/c^2}.$$

Here c is the speed of light.

☐ You should be able to show that if $v_{PB} = c$, then $v_{PA} = c$. If something moves with the speed of light in one frame, then it moves with the speed of light in all frames. You should also be able to show that if v_{PB} and v_{BA} are both much less than the speed of light, then the correct non-relativistic result is obtained.

NOTES:

Chapter 5
FORCE AND MOTION — I

In this, the most fundamental chapter in the mechanics section of the text, you start to study how objects influence the motion of each other. Be sure you understand the concepts of force and mass and pay particular attention to the relationship between the total force on an object and its acceleration.

Important Concepts

☐ Newton's first law   ☐ force of gravity (weight)

☐ inertial reference frame   ☐ tension (in a string)

☐ force   ☐ normal force

☐ mass   ☐ frictional force

☐ inertia   ☐ Newton's third law

☐ Newton's second law   ☐ free-body diagram

5–1 What Causes an Acceleration?

☐ The fundamental problem of mechanics is to find the acceleration of an object, given the object and its environment. The problem is split into two parts, connected by the idea of a force: the environment of an object exerts forces on the object and the total force on it causes it to accelerate. The first part of the problem is to find the total force exerted on the object, given the relevant properties of the object and its environment. The second part of the problem is to find the acceleration of the object, given the total force. In this chapter, we concentrate on the second part.

5–2 Newton's First Law

☐ **Newton's first law** helps us define **inertial reference frames.**
If the net force acting on a particle is zero, then the acceleration of the particle, as measured relative to an inertial frame, is also zero. Such a frame might be attached to a particle on which zero total force acts. Clearly the acceleration of the particle, as measured in that frame, is zero.

☐ The velocity of any inertial frame, measured relative to any other inertial frame, is a constant. Inertial reference frames do not accelerate relative to each other. The acceleration of a particle on which the total force is zero is itself zero in *every* inertial frame.

5–3 Force

☐ A **force** is a push or pull exerted by one object on another. It is measured, in principle, by applying the push or pull to the standard (1 kg) mass and measuring the acceleration of the standard mass. If SI units are used, the magnitudes of these quantities are numerically equal. Both are vectors in the same direction. That forces obey the laws of vector addition can be checked by simultaneously applying two forces in different directions and verifying that the result is the same as when the resultant of the forces is applied as a single force. All measurements must be made using an inertial reference frame.

☐ The SI unit of force is the newton and is abbreviated N. In terms of the SI base units, $1\,\text{N} = 1\,\text{kg} \cdot \text{m/s}^2$.

5–4 Mass

☐ The **mass** of an object is measured, in principle, by comparing the accelerations of the object and the standard mass when the same force is applied each to them. In particular, the mass of the object is given by $m = (1\,\text{kg})(a_0/a)$, where a is the magnitude of the acceleration of the object and

Chapter 5:   Force and Motion — I

a_0 is the magnitude of the acceleration of the mass standard. The accelerations must be measured using an inertial frame.

☐ An object with a small mass acquires a greater acceleration than an object with a large mass when the same force is applied to each of them. Mass is said to measure **inertia** or resistance to changes in motion.

☐ Mass is a scalar and is always positive. The mass of two objects in combination is the sum of the individual masses.

5–5 Newton's Second Law

☐ This is the central law of classical mechanics. It gives the relationship between the net force $\sum \mathbf{F}$ acting on an object and the acceleration $\mathbf{a}$ of the object:

$$\sum \mathbf{F} = m\mathbf{a},$$

where m is the mass of the object.

☐ The Newton's second law equation is a vector equation. It is equivalent to the three component equations

$$\sum F_x = ma_x, \quad \sum F_y = ma_y, \quad \text{and} \quad \sum F_z = ma_z.$$

☐ In these equations, $\sum \mathbf{F}$ is the *total* (or *net*) force acting on the object, the vector sum of all the individual forces. This means that in any given situation you must identify all the forces acting on the object and then sum them *vectorially*.

☐ Note that $\sum \mathbf{F} = 0$ implies $\mathbf{a} = 0$. If the resultant force vanishes, then the object does not accelerate; its velocity as observed in an inertial reference frame is constant in both magnitude and direction. The resultant force may vanish because no forces act on the object or because the forces that act sum to zero. For some situations, you may know that three forces act but are given only two of them. If you also know that the acceleration vanishes, you can solve $\mathbf{F}_1 + \mathbf{F}_2 + \mathbf{F}_3 = 0$ for the third force.

5–6 Some Particular Forces

☐ The **force of gravity** on an object is called the **weight** of the object and its magnitude is given by $W = mg$, where m is the mass of the object and g is the magnitude of the acceleration due to gravity at the position of the object. Near the surface of the Earth the direction of the weight is toward the center of the Earth.

☐ Be sure you understand that mass and weight are quite different concepts. Mass is a property of an object and does not change as the object is moved from place to place or even into outer space. It is a scalar. Weight, on the other hand, is a force. It varies as the object moves from place to place and vanishes when the object is far from all other objects, as in outer space. This is because **g**, not the mass, varies from place to place.

☐ Remember that the weight of an object is mg regardless of its acceleration. Weight is a force and, if appropriate, is included in the sum of all forces acting on the object. This sum equals m**a** and if other forces act, then **a** is different from **g**.

☐ The SI unit of weight is the newton.

☐ If a string with negligible mass connects two objects, it pulls on each with a force of the same magnitude, called the **tension** in the string. You may think of the string as simply transmitting a force from one object to the other; the situation is exactly the same if the objects are in contact and exert forces on each other. Strings pull, not push, along their lengths, so a string serves to define the direction of the force.

☐ A surface may exert a force on an object in contact with it. If the surface is frictionless, that force must be perpendicular to the surface. It is called a **normal force**. Unless adhesive glues the object to the surface a normal force can only push on the object. It must be directed away from the surface toward the interior of the object.

☐ If the surface is at rest or moving with constant velocity, the normal force adjusts until the component of the object's acceleration perpendicular to the surface vanishes. We often use this condition to solve for the normal force. Set the sum of the normal components of the forces acting on the object equal to zero. Since the normal force is one of these the resulting equation can be solved for it in terms of the normal components of the other forces.

☐ A surface may also exert a **frictional force** on an object in contact with it. This force is parallel to the surface.

5–7 Newton's Third Law

☐ Newton's third law tells us something about forces. If object A exerts a force $\mathbf{F}_{BA}$ on object B, then according to the third law, the force exerted by object B on object A is given by $\mathbf{F}_{AB} = -\mathbf{F}_{BA}$. Compared to the force of A on B, the force of B on A is the same in magnitude and opposite in direction. You should also be aware that these two forces are of the same type. That is, if object A exerts a *gravitational* force on B, then B exerts a *gravitational* force on A.

☐ The third law is useful in solving problems involving more than one object. If two objects exert forces on each other, we use the same symbol to represent their magnitudes and we remember the forces are in opposite directions when we write the second-law equations. In addition, we remember that the forces act on different objects. When we write Newton's second law for object A, one of the forces we include is the force of B on A, but emphatically *NOT* the force of A on B.

5–8 Applying Newton's Laws

☐ A definite procedure has been devised to solve Newton's second law problems. It ensures that you consider only one object at a time, reminds you to include all forces acting on the object you are considering, and guides you in writing Newton's second law in an appropriate form. Follow it closely.

1. Identify the object to be considered. It is usually the object on which the given forces act or about which a question is posed.

2. Represent the object by a dot on a diagram. Do not include the environment of the object since this is replaced by the forces it exerts on the object.

3. On the diagram draw arrows to represent the forces exerted by the environment on the object. Try to draw them in roughly the correct directions. The tail of each arrow should be at the dot. Label each arrow with an algebraic symbol to represent the magnitude of the force, regardless of whether a numerical value is given in the problem statement.

The hard part is getting all the forces. If appropriate, don't forget to include the weight of the object, the normal force of a surface on the object, and the forces of any strings or rods attached to the object. Carefully go over the sample problems in the text to see how to handle these forces.

Some students erroneously include forces that are not acting on the object. For each force you include, you should be able to point to something in the environment that is exerting the force. This simple procedure should prevent you from erroneously including a normal force, for example, when the object you are considering is not in contact with a surface.

4. Draw a coordinate system on the diagram. In principle, the placement and orientation of the coordinate system do not matter as far as obtaining the correct answer is concerned, but some choices reduce the work involved. If you can guess the direction of the acceleration, place one of the axes along that direction. The acceleration of an object sliding on a surface such as a table top or inclined plane,

for example, is parallel to the surface. Once the coordinate system is drawn, label the angle each force makes with a coordinate axis. This will be helpful in writing down the components of the forces later.

The diagram, with all forces shown but without the coordinate system, is called a **free-body diagram** (or a force diagram). We add the coordinate system to help us carry out the next step in the solution of the problem.

5.   Write the Newton's second law equation in component form: $\sum F_x = ma_x$, $\sum F_y = ma_y$, and, if necessary, $\sum F_z = ma_z$. The left sides of these equations should contain the appropriate components of the forces you drew on your diagram. You should be able to write the equations by inspection of your diagram. Use algebraic symbols to write them, not numbers; most problems give or ask for force magnitudes so you should usually write each force component as the product of a magnitude and the sine or cosine of an appropriate angle.

6.   If more than one object is important, as when two objects are connected by a string, you must carry out the steps above separately for each object. Then, you consider additional conditions. When two objects are connected by a string, for example, the magnitudes of their accelerations are the same. Be aware of these situations as you study the sample problems of the text.

7.   Identify which quantities are known and which are unknown; solve for the unknowns.

☐   As an example, a free-body diagram for a block of mass m on an inclined plane is shown below. The plane makes an angle θ with the horizontal. Two forces act on the block: the force of gravity (with magnitude mg) and the normal force of the plane (with magnitude N). The x axis is chosen to be down the plane and the y axis is chosen to be perpendicular to the plane, in the direction of the normal force. The x component of the force of gravity is $mg \sin \theta$ and the y component is $-mg \cos \theta$. The acceleration of the block is parallel to the x axis, so the second law equations are $mg \sin \theta = ma_x$ and $N - mg \cos \theta = 0$.

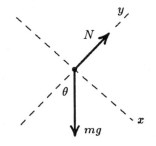

NOTES:

Chapter 6
FORCE AND MOTION — II

This chapter contains a great many applications of Newton's laws, with special emphasis on frictional and centripetal forces. Here's where your understanding of the fundamentals begins to pay off!

Important Concepts

☐ static friction
☐ coefficient of static friction
☐ kinetic friction
☐ coefficient of kinetic friction

☐ drag force
☐ drag coefficient
☐ terminal speed
☐ centripetal force

6–1 Friction

☐ A surface may exert a **frictional force** on an object that is in contact with it. Friction is unavoidable when the object is sliding on the surface, although lubricants and air films may make it small. Even when the object and surface are stationary with respect to each other, they exert frictional forces if other forces present would otherwise cause them to slide.

☐ A frictional force arises because microscopic welds form between protrusions from the two surfaces in contact. The macroscopic force of friction is actually the vector sum of a great many microscopic forces.

☐ Frictional forces are classified as being either static or kinetic. When the two objects are not moving relative to each other, the frictional force is labeled **static friction**. When the surfaces are moving relative to each other, the force is labeled **kinetic friction**.

□ A static frictional force is determined by the condition that the accelerations of the two surfaces are equal. Usually, but not always, an object rests on a surface (a table top or an inclined plane, for example) that is at rest. Then, the force of static friction on the object is just sufficient to hold it at rest.

□ All frictional forces, whether static or kinetic, are parallel to the surfaces in contact.

6–2 Properties of Friction

□ If an object is at rest on a motionless surface, the static frictional force is determined mathematically, via Newton's second law, by the condition that the component of acceleration parallel to the surface is zero. This condition is analogous to the condition used to determine the normal force. The difference is that the normal force of one object on another is perpendicular to the surface of contact while the frictional force is parallel to it.

□ The magnitude f_s of the force of static friction exerted by one surface on another must be less than a certain value, determined by the nature of the surfaces and by the magnitude of the normal force one surface exerts on the other. In particular, $f_s \leq \mu_s N$, where μ_s is the **coefficient of static friction** and N is the magnitude of the normal force. If the force of friction required to hold the surfaces at rest with respect to each other is greater than the maximum allowed, then the surfaces slide over each other and the frictional force is kinetic rather than static.

□ The magnitude of the force of kinetic friction is given by $f = \mu_k N$, where μ_k is the **coefficient of kinetic friction**. If the surface on which an object rests is motionless, the force of kinetic friction is opposite the velocity of the object.

□ The normal force that appears in the expressions for the force of kinetic friction and the maximum force of static friction must be computed for each situation using Newton's second law. As you know by now the magnitude of the normal force depends on the directions and magnitudes of other forces acting.

☐ Many problems of this chapter deal with frictional forces. Proceed as before: draw the free-body diagram and write down Newton's second law in component form, just as for any other problem. Use an algebraic symbol, f say, for the frictional force. You must now decide if the frictional force is static or kinetic. If static friction is involved, f is probably an unknown but the acceleration is known or is related to other known quantities in the problem. If the object is at rest on a stationary surface, its acceleration is zero. If it is at rest relative to an accelerating surface, its acceleration is the same as that of the surface. Kinetic friction is involved if one surface is sliding on the other. Then, the magnitude of the frictional force is given by $\mu_k N$.

☐ If you do not know that the object is at rest relative to the surface, assume it is and use Newton's second law, with the acceleration of the object equal to the acceleration of the surface, to calculate both the force of static friction f_{rest} that will hold it at rest and the normal force N. Compare f_{rest} with $\mu_s N$. If $f_{\text{rest}} < \mu_s N$, the object remains at rest relative to the surface and the force of friction has the value you computed. That is, $f = f_{\text{rest}}$. If $f_{\text{rest}} > \mu_s N$, then the object moves relative to the surface. Go back to the second law equations and set $f = \mu_k N$, then solve for the acceleration.

6–3   The Drag Force and Terminal Speed

☐ When an object moves in a fluid, such as air, the fluid exerts a **drag force** on it. When the relative speed of the object and fluid is so great that the fluid flow around the object is turbulent, the force of the fluid on the object is proportional to the square of the relative speed. In fact, the magnitude of the drag force is given by: $D = \frac{1}{2}C\rho A v^2$, where v is the relative speed of the object, A is the effective cross-sectional area of the object, ρ is the mass density of the fluid, and

C is a drag coefficient. The direction of the drag force is opposite the direction of the relative velocity of the object.

☐ When an object falls in a fluid, it approaches a constant speed, called the **terminal speed**. You can use Newton's second law to find a value for this speed. The force of gravity is mg, down, and the drag force is $\frac{1}{2}C\rho Av^2$, up. At terminal speed these sum to zero. Thus,

$$v_t = \sqrt{\frac{2mg}{C\rho A}}\,.$$

The larger the combination $C\rho A$ the smaller the terminal speed and the shorter the time taken to reach that speed from rest.

☐ You should realize that when the object is dropped from rest its acceleration is g at first. As it picks up speed the drag force increases, thereby reducing the acceleration. The object continues to gain speed but at a lesser rate. At terminal speed its acceleration vanishes. From then on the acceleration remains zero, the speed does not change, and the drag force remains constant.

6–4 Uniform Circular Motion

☐ An object in uniform circular motion has a non-zero acceleration because the direction of its velocity changes with time. A force must be applied to the object in order to produce its acceleration. If m is the mass of the object, then the applied force must have magnitude $F = mv^2/r$, where r is the radius of the orbit and v is the speed of the object. The force is directed toward the center of the circular orbit and, because of its direction, is called a **centripetal force**.

☐ Acquire the habit of pointing out to yourself the object in the environment that exerts the force. It might be a string, for example. If you are sitting in a car without a restraining belt and the car rounds a curve, the force that pulls you around the curve with the car is provided by friction between you and the car seat. If this force is not great enough, you slide toward the outside of the curve; you are going to a larger-radius path. If the force is zero (a very slippery seat), you will travel in a straight line as the seat moves out from under you around the curve.

☐ If a force **F** is applied to an object of mass m traveling at speed v and the force is maintained perpendicular to the velocity, then the object will travel with constant speed in a circle of radius $r = mv^2/F$. If the magnitude of the force is decreased, the radius of the path will increase.

☐ Uniform circular motion problems are solved in much the same way as any other second law problem. Draw a free-body diagram. Place the coordinate system so one of the axes is in the direction of the acceleration, pointing from the object toward the center of its orbit. For most problems, you will want to substitute v^2/r for the magnitude of the acceleration. Here v is the speed of the object and r is the radius of its orbit.

6–5 The Forces of Nature

☐ There are only a small number of fundamental forces in nature:

1. Gravitational force. All objects with mass attract each other gravitationally. The force is very weak at the atomic level, where the masses are small, but this type force is important at the macroscopic level.

2. Electromagnetic force. Electric and magnetic forces are different manifestations of this force. At the macroscopic level all contact forces are electromagnetic in nature.

3. Strong force. This binds together the constituent particles of atomic nuclei. It is extremely strong when nuclear particles are close together, as within a nucleus, but its range is extremely short.

4. Weak force. This is another short range nucle
It is responsible for the spontaneous decay of cer
clear particles.

☐ Physicists believe that these forces are actually all
manifestations of the same more fundamental fo
weak and electromagnetic forces have been sho
different aspects of the same force, called the ele
force. Earlier, electric and magnetic forces were
be different aspects of the same force, now called
tromagnetic force.

NOTES:

Chapter 7
KINETIC ENERGY AND WORK

The central concept of this chapter is the idea of work. You should understand the definition of work, you should learn how to calculate the work
done by forces in various situations, and you should learn how work is
related to the change in the kinetic energy of a particle (the work-kinetic
energy theorem).

Important Concepts

☐ kinetic energy

☐ work-kinetic energy theorem

☐ work

☐ work done by gravity

☐ force of an ideal spring

☐ spring constant

☐ work done by an ideal spring

☐ power

7–1   Kinetic Energy

☐   The **kinetic energy** of a particle with mass m and speed v is
defined by

$$K = \tfrac{1}{2}mv^2 \, .$$

For a particle moving in the xy plane, $K = \frac{1}{2}m(v_x^2 + v_y^2)$,
where v_x and v_y are its velocity components. The two terms
of the expression are *not* components of a vector. Kinetic
energy is a scalar and so does not have a direction associated with it.

☐   The SI unit of kinetic energy is the joule (abbreviated J).
In terms of SI base units, $1\,\text{J} = 1\,\text{kg} \cdot \text{m}^2/\text{s}^2$. A unit of
energy called an electron volt is often used when dealing
with atomic phenomena. In terms of the SI unit, $1\,\text{eV}$ is
$1.60 \times 10^{-19}\,\text{J}$.

7–2 Work

☐ **Work** is energy that is being transferred to or from an object by a force acting on the object. If the object has no other form of energy, the work done on it changes its kinetic energy.

7–3 Work and Kinetic Energy

☐ The **work-kinetic energy theorem** is: during any interval the *total* work done by *all* forces acting on a particle equals the change in the particle's kinetic energy. Let W be the total work done on a particle during some interval. If K_i is the initial kinetic energy and K_f is the final kinetic energy for that interval, then

$$W = K_f - K_i .$$

Be sure you perform the subtraction in the correct order.

☐ If the total work is negative, then the kinetic energy and speed of the particle both decrease; if it is positive, then the kinetic energy and speed both increase; if it is zero, then the kinetic energy and speed do not change.

☐ The work-kinetic energy theorem is a direct result of Newton's second law. The derivation is given in Section 7–5 of the text.

☐ The work-kinetic energy theorem is valid only for a single particle or an object that can be treated as a particle. If the object becomes distorted or its orientation changes, then the theorem cannot be applied directly to the object as a whole.

☐ If a constant force **F** acts on a particle as it moves through a displacement **d**, the force does work $W = \mathbf{F} \cdot \mathbf{d} = Fd \cos \phi$, where ϕ is the angle between the force and the displacement when they are drawn with their tails at the same point. If the particle has displacement Δx, along the x axis, and the force has x component F_x, then the expression for the work can be written $W = F_x \Delta x$.

☐ Work is a scalar. It does not have a direction associated with it. If several forces act on the particle, the total work done by all the forces is the algebraic sum of the individual works.

☐ Work is positive if the angle ϕ is less than 90°; it is negative if ϕ is greater than 90°. A force that is perpendicular to the displacement ($\phi = 0$) does no work. A person carrying a box horizontally with constant velocity does no work on the box. The normal force of a stationary surface on a sliding crate does no work on the crate, no matter what the orientation of the surface. A force also does no work if the displacement is zero. A person holding a book at rest does no work on the book.

☐ The SI unit of work is the same as that for energy, the joule.

7–4 Work Done by Weight

☐ As an object of mass m moves from height y_i to height y_f near the Earth's surface, gravity does work $W = -mg(y_f - y_i)$. This expression is valid if $y_f > y_i$ (the object rises during the interval) or if $y_f < y_i$ (the object is falling during the interval). On the way down the sign of the work done by gravity is positive while on the way up it is negative.

☐ The work done by gravity is the same no matter what path is taken between the initial and final points. In fact, all that counts is the initial and final altitudes. The two positions need not have the same horizontal coordinate.

☐ The expression for the work done by gravity is valid even if the object experiences air resistance. If air resistance is present, this expression does not, of course, give the total work done by all forces.

7–5 Work Done by a Variable Force

☐ Suppose a particle moves along a straight line (the x axis) and is subjected to a force $\mathbf{F}$ that is parallel to that line and is not constant. Then, the work done by the force as the particle moves from x_i to x_f is given by the integral

$$W = \int_{x_i}^{x_f} F(x)\, dx\,.$$

To evaluate the integral, the x component of the force must be known as a function of the particle coordinate x. The work done by the force is the area under the graph of F vs. x.

☐ If a particle moves in two or three dimensions, subject to a variable force **F**, the work done by the force on the particle as it moves from $\mathbf{r}_i$ to $\mathbf{r}_f$ is given by the integral

$$W = \int_{\mathbf{r}_i}^{\mathbf{r}_f} \mathbf{F} \cdot d\mathbf{s}\,.$$

Here d**s** is an infinitesimal displacement along the path of the particle. This expression is the general definition of work. The expressions given above for motion in one dimension can be derived from it.

7–6 Work Done by a Spring Force

☐ The force exerted by an ideal spring on an object is a variable force. Its direction depends on whether the spring is extended or compressed and its magnitude depends on the amount of extension or compression. Assume the spring is along the x axis with one end fixed and the other attached to the object, as shown below. When the object is at $x = 0$, the spring is neither extended or compressed. This is the equilibrium configuration. When the object is at any coordinate x, the force exerted by the spring is given by $F = -kx$. Here k is the **spring constant**, a property of the spring: the stiffer the spring, the greater the spring constant.

Chapter 7:   Kinetic Energy and Work

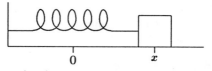

- [ ] When the spring is extended, the force it exerts is negative; when the spring is compressed the force is positive. In either case, the force tends to push the object toward the equilibrium point ($x = 0$). The force is called a *restoring force*.

- [ ] The spring constant of an ideal spring can be measured by applying a force F to the mass and measuring the elongation x of the spring with the mass at rest. In terms of F, x, and k, the net force on the mass is given by $F - kx$ and since the mass is at rest this must be zero. Thus, $k = F/x$. The SI unit of a spring constant is N/m.

- [ ] As the object moves from an initial coordinate x_i to a final coordinate x_f the work done by the spring is

$$W = \int_{x_i}^{x_f} -kx \, \mathrm{d}x = -\tfrac{1}{2}k \left(x_f^2 - x_i^2 \right) .$$

- [ ] The spring does positive work whenever the object moves toward the equilibrium point from either side and does negative work whenever the object moves away from the equilibrium point.

7–7 Power

- [ ] The **power** associated with a force is the rate at which it does work. Let the function $W(t)$ represent the work done by a force from time 0 to time t. Then, the instantaneous power delivered by the force is given by

$$P = \frac{\mathrm{d}W}{\mathrm{d}t} .$$

- [ ] The SI unit of power is the watt (abbreviated W): $1 \, \text{W} = 1 \, \text{J/s}$.

☐ Suppose that at some instant of time a particle is moving with velocity **v** and is acted on by a force **F**. Then, the power delivered to the particle by the force is given by $P = \mathbf{F} \cdot \mathbf{v}$.

7–8 Kinetic Energy at High Speeds

☐ When an object is moving with a speed near the speed of light c, then the relativistic definition of kinetic energy must be used. If m is the mass and v is the speed, then the kinetic energy is defined by

$$K = mc^2 \left(\frac{1}{\sqrt{1 - (v/c)^2}} - 1 \right) .$$

At speeds that are much less than the speed of light this expression reduces to $K = \frac{1}{2}mv^2$.

7–9 Reference Frames

☐ If two observers use different inertial frames (moving relative to each other) to measure the change in kinetic energy of an object over some interval, they will obtain different values. They will also obtain different values for the total work done on the object, but the work-kinetic energy theorem will still be valid for both frames. Each will conclude that the change in kinetic energy equals the total work done. This is an example of the **principle of invariance**, which states that the laws of physics have the same form in all inertial frames.

NOTES:

Chapter 8
POTENTIAL ENERGY
AND CONSERVATION OF ENERGY

The closely related concepts of conservative force and potential energy are central to this chapter. Pay close attention to their definitions. If *all* forces exerted by objects in a system on each other are conservative and no net work is done on them by outside agents, then the mechanical energy of the system (the sum of the kinetic and potential energies) does not change. When non-conservative forces act, another energy, called the internal energy of the system, must be included in the sum for the principle of energy conservation to hold. When external agents do work on the system, its energy is not conserved.

Important Concepts

☐ potential energy
☐ gravitational potential energy
☐ elastic (spring) potential energy
☐ mechanical energy
☐ conservative force
☐ nonconservative force
☐ potential energy curve
☐ turning point

☐ equilibrium points
☐ stable equilibrium
☐ unstable equilibrium
☐ neutral equilibrium
☐ internal energy
☐ conservation of energy
☐ mass energy
☐ energy quantization

8–1 Potential Energy

☐ **Potential energy** is an energy associated with the configuration of two or more interacting objects; that is, it depends primarily on the positions of the objects. For example, the potential energy associated with the gravitational interaction between the Earth and another object depends on the separation of the Earth and the object. The greater the separation the greater the gravitational potential energy.

☐ Potential energy may be thought of as stored in the system. For example, the gravitational potential energy of a ball and the Earth is stored in the Earth-ball system.

8–2 Path Independence of Conservative Forces

☐ A force is **conservative** if the work that it does is zero as the object that exerts it and the object on which it is exerted move from any configuration back to the same configuration, regardless of the path taken by the objects. One direct consequence is that the work done by a conservative force acting between objects of a system, as the system goes from one configuration to another, does not depend on the paths taken by objects of the system.

☐ The force of gravity and the force of a spring are conservative forces. Friction is a nonconservative force.

☐ A potential energy is associated with a conservative force. A potential energy cannot be associated with a nonconservative force.

8–3 Determining Potential Energy Values

☐ As the objects of a system move the forces they exert on each other may do work. If any of these forces is conservative, the change in the potential energy associated with it is the negative of the work it does: $\Delta U = -W$.

☐ Only *changes* in the potential energy are physically meaningful. Usually the potential energy is chosen to have some value for a particular reference configuration of the system. Then, the work done by forces of the system as the system goes from the reference configuration to another configuration is computed. If $U(x_0)$ represents the potential energy in the reference configuration and $U(x)$ represents the potential energy in the other configuration, then

$$U(x) = U(x_0) - W,$$

where W is the work. $U(x_0)$ is often taken to be zero.

☐ Potential energy is a scalar. If two or more conservative forces act, the total potential energy is simply the algebraic sum of the individual potential energies. The total potential energy of the system is the sum of the potential energies associated with all the conservative forces of mutual interaction.

☐ The SI unit for potential energy is the joule.

☐ The gravitational potential energy of a system consisting of the Earth and an object of mass m, near its surface, is given by

$$U(y) = 0 - \int_0^y (-mg)\,\mathrm{d}y = mgy\,,$$

where the y axis is taken to be positive in the upward direction. U was chosen to be zero at $y = 0$. Gravitational potential energy is often ascribed to the object alone, but it is actually a property of the Earth-mass system.

☐ Gravitational potential energy depends only on the altitude of the object above the surface of the Earth, even if the object moves horizontally as well as vertically.

☐ If the altitude of an object above the Earth is increased, the work done by gravity is negative and the potential energy change is positive. If the altitude is decreased, the work done by gravity is positive and the potential energy change is negative.

☐ Remember that the force of an ideal spring on an object attached to one end is given by $F = -kx$, where k is the spring constant and x is the coordinate of the object, measured from an origin at the point where the spring is neither elongated nor compressed. This force always pulls the object toward the equilibrium point ($x = 0$).

☐ The potential energy of a system consisting of an object attached to an ideal spring with spring constant k is given by

$$U(x) = 0 - \int_0^x (-kx)\,\mathrm{d}x = \tfrac{1}{2}kx^2\,,$$

where U was chosen to be zero at $x = 0$. The greater the elongation or compression of the spring the greater the potential energy.

☐ An external agent can increase the spring potential energy by compressing or extending the spring. When a spring is elongated, the work done by the spring is negative and the change in the spring potential energy is positive. When the spring is compressed, the work done by the spring is again negative and the change in the spring potential energy is again positive. This stored potential energy is converted to kinetic energy when the spring is released.

8–4 Conservation of Mechanical Energy

☐ The **mechanical energy** of a system is the sum of its potential and kinetic energies:

$$E = K + U.$$

If the mechanical energy remains constant as objects in the system move, it is said to be **conserved**. The mechanical energy of a system is conserved if a potential energy is associated with each of the forces between objects of the system and no external agents do work on the system. This follows directly from the work-kinetic energy theorem and the definition of potential energy.

☐ If a block on a horizontal frictionless surface is attached to an ideal spring and the block is pulled aside and released, it oscillates back and forth, repeatedly coming back to its original position. As the block moves toward the equilibrium point ($x = 0$) it speeds up and the spring becomes less extended or compressed; as it moves away from the equilibrium point it slows down and the spring becomes more extended or compressed. In the first case, the block gains kinetic energy and the spring loses potential energy; in the second case, the block loses kinetic energy and the spring gains potential energy. The sum of the potential energy stored in the spring and the kinetic energy of the block remains constant throughout; the mechanical energy of the spring-block system is conserved. $\frac{1}{2}mv^2 + \frac{1}{2}kx^2 = $ constant can be used to solve for one of the quantities that appear in it.

☐ If a ball is thrown upward, the force of gravity slows it down; its kinetic energy decreases but the gravitational potential energy of the Earth-ball system increases. Similarly, as the ball falls its kinetic energy increases and the gravitational potential energy of the Earth-ball system decreases. If the gravitational force is the only force acting on the ball, the sum of the potential and kinetic energies remains constant during the flight of the ball; the mechanical energy of the Earth-ball system is conserved. $\frac{1}{2}mv^2 + mgy = $ constant can be used to solve for one of the quantities that appear in it.

☐ If frictional forces act, the mechanical energy of a system decreases. Mechanical energy is said to be **dissipated**.

8–5 Reading a Potential Energy Curve

☐ If the potential energy function is known, the force can be computed by evaluating its derivatives with respect to coordinates. If $U(x)$ is the potential energy for an object acted on by a conservative force F, then

$$F(x) = -\frac{dU(x)}{dx}.$$

On a graph of the potential energy as a function of x the force is the negative of the slope of the line that is tangent to the curve at the coordinate of the particle. A positive slope indicates a force in the negative x direction and a negative slope indicates a force in the positive x direction.

☐ When dealing with potential energy curves, assume that the agent exerting the force is stationary, so only the object whose motion is being considered has kinetic energy.

☐ A potential energy curve is shown below for an object that moves along the x axis with constant mechanical energy E, represented by a horizontal dotted line. The potential and kinetic energies when the object is at x_1 are indicated by arrows. The object cannot move to the left of x_{min} or to the right of x_{max} because the potential energy is greater than the mechanical energy in those regions. The kinetic energy and the velocity are zero at x_{min} and x_{max}. When the object is at x_{min}, the force acting on it is positive (the slope of the curve is negative) and the object is pushed to the right. When it is at x_{min}, the force acting on it is negative and the object is pushed to the left. x_{min} and x_{max} are called the **turning points** of the motion.

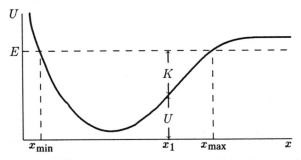

☐ Suppose the object starts at x_1 and travels in the negative x direction. Its speed increases until it gets to the coordinate corresponding to the minimum in the potential energy curve; then it slows. When it reaches x_{min}, its velocity is zero but a positive force acts on it and it starts moving in the positive x direction. Its speed increases, then decreases until it gets to x_{max}, where its velocity is zero. It then starts moving back again and the motion is repeated. If the energy is decreased, it oscillates over a narrower range of coordinates.

☐ **Equilibrium points** are places where the force vanishes. The slope of the potential energy curve is zero at these points. When an object is released from rest near a point where the curve has a minimum (a point of **stable equilibrium**), its subsequent motion is toward the equilibrium point. When an object is released from rest near a point where the curve has a maximum (a point of **unstable equilibrium**), its subsequent motion is away from the equilibrium point. When an object is released from rest in a region where the curve is horizontal (a region of **neutral equilibrium**), it remains at rest.

8–6 Work Done by Nonconservative Forces

☐ If external agents do work on the objects of a system, the mechanical energy of the system is not conserved. Then, the energy equation becomes

$$\Delta K + \Delta U = W_{\text{app}},$$

where W_{app} is the total work done *on* objects of the system by outside agents. An external force doing work on objects in a system transfers energy to or from the system. For example, the work done by a person carrying a crate up some stairs increases the mechanical energy of the crate-Earth system. If the speed of the crate does not change, the increase in energy appears as potential energy; the crate is further away from the Earth than previously. If the crate's speed increases, both the potential and kinetic energies increase.

☐ If a constant frictional force **f** acts on an object as it moves through the displacement **d**, then $\mathbf{f} \cdot \mathbf{d}$ is used, along with the work done by any other forces, to determine the change in the kinetic energy of the block. This conclusion follows directly from Newton's second law: **f** is included with all other forces to determine the total force **F** acting on the block and the integral $\int \mathbf{F} \cdot d\mathbf{s}$ gives ΔK.

☐ Because a force of friction **f** is actually the sum of a large number of forces, acting at the welds that form between two surfaces, and the welds move through displacements that are different from the displacement of the object, the work done by friction is *not* given by $\mathbf{f} \cdot \mathbf{d}$, where $\mathbf{d}$ is a displacement of the object. In fact, the actual work done by friction transfers internal energy between the object and the surface on which it slides.

8–7 Conservation of Energy

☐ If a nonconservative force, such as friction, acts on objects of a system, the internal energy of the objects must be included in the total energy of the system.

☐ The **internal energy** E_{int} of an object is the sum of the kinetic and potential energies associated with random motions of the particles making up the object and with their mutual interactions, as distinct from the motion of the object as a whole or its interaction with other objects. E_{int} must be included whenever work done on the object results in changes in the motions of the constituent particles that are different from changes in the motion of the object as a whole. Nonconservative forces, such as friction, always change the internal energy of the object on which they act because the welds between the surfaces deform and the microscopic surface structure changes as time goes on.

☐ When internal energy is included, the energy of a system of objects becomes

$$E_{tot} = K + U + E_{int}.$$

Here K is the total kinetic energy of the objects, U is the total potential energy of their interactions with each other, and E_{int} is the total internal energy.

☐ Whenever mechanical energy is not conserved, physicists have been able to restore the principle of energy conservation by defining other forms of energy and including them in the energy balance. The total is conserved if no external agents do work on the system. The energy may change form and may be transferred from one object to another in the system but the total remains constant.

☐ For an object sliding on a surface, an isolated system (no external forces doing work) consists of the surface, the Earth, and the object. E_{int} is then the sum of the internal energies of the object and the surface.

☐ If external forces do work W on objects in the system, then the total energy of the system, including its internal energy, changes:

$$\Delta E_{tot} = W .$$

8–8 Mass and Energy

☐ Nuclear reactions might change the mass of a system. A large number of new particles often emerge from the collision of two protons at high kinetic energy, for example. An energy, called the **mass energy** is associated with mass and in a nuclear reaction it may be converted to kinetic or potential energy. Other forms of energy may also be converted to mass energy.

☐ A particle of mass m has a mass energy that is given by $E = mc^2$, where c is the speed of light. The negative of the change in the total mass energy is called the Q of a reaction:

$$Q = -\Delta m\, c^2 .$$

8–9 Quantized Energy

☐ The energy of a system is quantized. That is, it may have only certain discrete values and other values never occur. Quantum mechanics correctly predicts the allowed energy values for any given system. For macroscopic systems, the separation between allowed energy values is too small to be detected, but for atomic and subatomic systems energy quantization has profound effects. For example, Fig. 8–17 shows some of the allowed energy values for a typical atom.

☐ The energy of an atomic system can change from one allowed value to another with the emission or absorption of electromagnetic radiation (light, for example). The energy of the radiation must match exactly the difference in two allowed energies.

NOTES:

Chapter 9
SYSTEMS OF PARTICLES

This chapter is about systems of more than one particle. The center of mass is important for describing the motion of the system as a whole, and its velocity is closely related to the total momentum of the system. External forces acting on the system accelerate the center of mass and when the net external force vanishes the velocity of the center of mass is constant. The total momentum of the system is then conserved.

Important Concepts

☐ center of mass

☐ Newton's second law
   for the center of mass

☐ linear momentum

☐ conservation of
   linear momentum

☐ kinetic energy of
   the center of mass

9–1   A Special Point

☐ The motion of an object might be quite complicated: the object might rotate and deform as it moves. Nevertheless, the motion of a special point, called the **center of mass** of the object, is much less complicated than the motion of other parts of the object. It moves like a point particle that is subjected only to the external forces applied to the object. No matter how a thrown object twists and turns, for example, its center of mass follows the parabolic trajectory of an ideal projectile.

9–2   The Center of Mass

☐ The coordinates of the center of mass of a collection of particles are given by

$$x_{cm} = \frac{1}{M} \sum_{i=1}^{N} m_i x_i \,, \qquad y_{cm} = \frac{1}{M} \sum_{i=1}^{N} m_i y_i \,,$$

and

$$z_{cm} = \frac{1}{M} \sum_{i=1}^{N} m_i z_i \,,$$

where M is the total mass of the system, m_i is the mass of particle i, and x_i, y_i, z_i are its coordinates. In vector notation, the position vector of the center of mass is given by

$$\mathbf{r}_{cm} = \frac{1}{M} \sum_{i=1}^{N} m_i \mathbf{r}_i \,,$$

where $\mathbf{r}_i$ is the position vector of particle i. The center of mass is not necessarily at the position of any particle in the system.

☐ For a *continuous* distribution of mass, the coordinates of the center of mass are computed by dividing the object into a large number of infinitesimal regions, each with mass dm. Each region is treated like a particle and the contributions are summed. The results are written as the integrals

$$x_{cm} = \frac{1}{M} \int x \, dm \,, \qquad y_{cm} = \frac{1}{M} \int y \, dm \,,$$

and

$$z_{cm} = \frac{1}{M} \int z \, dm \,.$$

☐ If the mass is distributed symmetrically about some point or line, then the center of mass is at that point or on that line. For example, the center of mass of a uniform spherical shell is at its center, the center of mass of a uniform sphere is at its center, the center of mass of a uniform cylinder is at the midpoint of its central axis, the center of mass of a uniform rectangular plate is at its center, and the center of mass of a uniform triangular plate is at the intersection of the lines from the vertices to the midpoints of the opposite sides and halfway through its thickness.

□ Sometimes an object with a complicated shape can be divided into parts such that the center of mass of each part can be found easily. Then, each part is replaced by a particle with mass equal to the mass of the part, positioned at the center of mass of the part. The center of mass of the whole object is at the position of the center of mass of these particles. This idea can also be used to find the center of mass of an object with a hole.

9–3 Newton's Second Law for a System of Particles

□ As the particles of a system move, the coordinates of the center of mass might change. Consider a system of discrete particles and differentiate the expression for the center of mass position vector to find the velocity of the center of mass in terms of the individual particle velocities:

$$\mathbf{v}_{\text{cm}} = \frac{1}{M} \sum_{i=1}^{N} m_i \mathbf{v}_i \, ,$$

where $\mathbf{v}_i$ is the velocity of particle i.

□ As the particles of a system accelerate the center of mass might accelerate. Differentiate the expression for the velocity of the center of mass to find the acceleration of the center of mass in terms of the accelerations of the individual particles:

$$\mathbf{a}_{\text{cm}} = \frac{1}{M} \sum_{i=1}^{N} m_i \mathbf{a}_i \, ,$$

where $\mathbf{a}_i$ is the acceleration of particle i.

□ Since each particle of a system obeys Newton's second law, $\sum \mathbf{F}_i = M \mathbf{a}_{\text{cm}}$, where $\mathbf{F}_i$ is the vector sum of all forces on particle i, some of which might be due to other particles in the system and some of which might be due to the environment of the system. The sum $\sum \mathbf{F}_i$ is the vector sum of all forces acting on all particles of the system but, because Newton's third law is valid, it reduces to $\sum \mathbf{F}_{ext}$, the sum over all *external* forces acting on particles of the system, i.e. those due to the environment of the system. If two particles of the system exert forces on each other, the two forces are equal in magnitude and opposite in direction. Both occur in the sum $\sum \mathbf{F}_i$ and so cancel each other. Thus, the center of mass obeys a Newton's second law equation:

$$\sum \mathbf{F}_{ext} = M\mathbf{a}_{cm} .$$

The mass that appears in this law is the total mass of the system (the sum of the masses of the individual particles) and the force that appears is the vector sum of all *external* forces acting on all particles of the system.

☐ If the total external force acting on a system is zero, then the acceleration of the center of mass is zero and the velocity of the center of mass is constant. If the center of mass of the system is initially at rest and the total external force acting on the system is zero, then the velocity of the center of mass is always zero and the center of mass remains at its initial position.

☐ These statements are true no matter what forces the particles of the system exert on each other. The particles might, for example, be fragments that are blown apart in an explosion, they might be objects that collide violently, or they might be objects that interact with each other from afar via gravitational or electrical forces.

9–4 Linear Momentum

☐ The momentum of a particle of mass m, moving with a velocity $\mathbf{v}$ (assumed to be much less than the speed of light) is given by $\mathbf{p} = m\mathbf{v}$. Momentum is a vector. Its SI units are $kg \cdot m/s$.

☐ In terms of momentum, Newton's second law for a particle is $\sum \mathbf{F} = d\mathbf{p}/dt$, where $\sum \mathbf{F}$ is the total force acting on the particle.

☐ If the particle speed is close to the speed of light, a relativistically correct expression must be used. The momentum of a particle of mass m moving with velocity $\mathbf{v}$ is then given by

$$\mathbf{p} = \frac{m\mathbf{v}}{\sqrt{1 - (v/c)^2}},$$

where c is the speed of light.

9–5 The Linear Momentum of a System of Particles

☐ The total momentum $\mathbf{P}$ of a system of particles is the vector sum of the individual momenta. Since, for non-relativistic particles, $\mathbf{P} = m_1\mathbf{v}_1 + m_2\mathbf{v}_2 + \ldots$, the total momentum is given by $\mathbf{P} = M\mathbf{v}_{cm}$, where M is the total mass of the system. That is, the total momentum of the system is identical to the momentum of a single particle with mass equal to the total mass of the system, moving with a velocity equal to the velocity of the center of mass.

☐ Newtons' second law for a system of particles can be written in terms of the total momentum: $\sum \mathbf{F}_{ext} = d\mathbf{P}/dt$.

9–6 Conservation of Linear Momentum

☐ The principle of momentum conservation is: If the total external force acting on a system of particles vanishes, then the total momentum of the system is constant. Total momentum is said to be conserved. Since momentum is a vector this principle is equivalent to the three equations: $P_x =$ constant if $\sum F_{ext\,x} = 0$, $P_y =$ constant if $\sum F_{ext\,y} = 0$, and $P_z =$ constant if $\sum F_{ext\,z} = 0$. One component of momentum might be conserved even if the others are not.

☐ If the momentum of a system is conserved, then the acceleration of the center of mass of the system is zero and the velocity of the center of mass is constant. If the momentum of the system is zero, then its center of mass does not move.

9–7 Systems with Varying Mass: A Rocket

☐ The conservation of momentum principle can be used to find an equation for the acceleration of a rocket if the system includes both the rocket and the fuel it expels. Any external forces, such as gravity, are assumed to be negligible.

☐ Suppose that at time t a rocket with mass M (including fuel) is moving with velocity v. It will eject fuel of mass $-\Delta M$ (a positive quantity) in time Δt. Originally the fuel is traveling with the rocket at velocity v but after it is ejected it has velocity U and the rocket has velocity $v + \Delta v$. Carefully note that v, $v + \Delta v$, and U are all measured relative to an inertial frame and that ΔM is taken to be negative, so the mass of the rocket after ejection is $M + \Delta M$. Also note that the rocket exerts a force on the fuel to eject it and the fuel exerts a force on the rocket of the same magnitude but in the opposite direction.

☐ The rocket and the ejected fuel form a constant mass system on which no external forces act, so the total momentum Mv before the fuel is ejected is equal to the total momentum $-dM\,U + (M + dM)(v + dv)$ after the fuel is ejected. Divide by dt and take the limit as dt becomes vanishingly small. Products of two infinitesimals vanish and the result is $(U - v)\,dM/dt = M\,dv/dt$.

☐ This expression is usually written in terms of the fuel speed *relative to the rocket*. Assume the direction of fuel ejection is opposite the direction of travel of the rocket and let u be the relative speed. Then, $u = v - U$ and $-u\,dM/dt = M\,dv/dt$. The rate at which the rocket loses mass is $R = -dM/dt$ and dv/dt is the acceleration a of the rocket, so the rocket equation becomes $Ru = Ma$. This equation is used to find the acceleration of the rocket in terms of the rate of fuel consumption. The quantity Ru is called the **thrust** of the rocket engine.

☐ Integrate $dv = -u\,dM/M$ from some initial velocity v_i and mass M_i to some final velocity v_f and mass M_f to obtain

$$v_f - v_i = u \ln \frac{M_i}{M_f} .$$

This expression gives the change in the rocket's velocity in terms of the initial and final masses. $M_i - M_f$ is the mass of the expended fuel.

9–8 External Forces and Internal Energy Changes

☐ The total kinetic energy of an object can be divided into two parts. One is associated with motion of the object as a whole and is called the **kinetic energy of the center of mass**. The other is associated with the motions of the various particles that make up the object, measured relative to the center of mass. It and the potential energy associated with mutual interactions of particles in the system make up the internal energy of the object.

☐ For a system of particles with total mass M and center of mass velocity $\mathbf{v}_{cm}$, the kinetic energy of the center of mass of the system is given by $K_{cm} = \frac{1}{2} M v_{cm}^2$. Changes in this kinetic energy are related to the total external force acting on the system and the displacement of the center of mass. For purposes of calculating ΔK_{cm}, the system is replaced by a single particle of mass M, located at the position of the center of mass and acted on by the total external force $\sum \mathbf{F}_{ext}$. The change in the kinetic energy of the center of mass then equals the work W_{cm} done by this force on this "particle".

☐ Suppose the center of mass of a system moves from x_i to x_f along the x axis, during which motion the total external force is F_{ext}, a constant. The work done by this force on the "particle" that replaces the system is given by $W_{cm} = F(x_f - x_i)$ and the change in the kinetic energy of the center of mass of the system is given by $\Delta K_{cm} = F(x_f - x_i)$.

☐ W_{cm} is not necessarily the actual work done by the resultant external force. In some cases, the actual work done by $\sum \mathbf{F}_{ext}$ may change the internal energy of the system as well as change the kinetic energy of the center of mass. In other cases, $\sum \mathbf{F}_{ext}$ may not actually do any work but W_{cm} is not zero and the kinetic energy changes. This occurs, for example when the point of application of an external force does not move but the center of mass does. Then, the entire change in the kinetic energy of the center of mass arises from a change in the internal energy of the object. If you are asked for a change in energy, use the energy equation $\Delta K_{cm} + \Delta U + \Delta E_{int} = W$, where W is the actual work done by external forces. If you are asked about a change in the kinetic energy of the center of mass, use $W_{cm} = \Delta K_{cm}$.

NOTES:

Chapter 10
COLLISIONS

Here the principle of momentum conservation is applied to collisions between objects. Pay special attention to the role played by the impulses the objects exert on each other. Also take careful notice of when energy is conserved in a collision and when it is not.

Important Concepts

☐ collision   ☐ inelastic collision

☐ impulse   ☐ completely inelastic collision

☐ elastic collision

10–1   What is a Collision?

☐   In a collision, two objects exert relatively strong forces on each other for a relatively short time. A collision takes place during a well-defined time interval: the objects do not interact with each other before or after.

10–2   Impulse and Linear Momentum

☐   During a collision each object exerts a force on the other. What is important is not the force alone or its duration alone but a combination called the **impulse** of the force. If one body acts on the other with a force $\mathbf{F}(t)$ for a time interval from t_i to t_f, the impulse of the force is given by

$$\mathbf{J} = \int_{t_i}^{t_f} \mathbf{F}(t)\, dt\,.$$

Don't confuse impulse and work. Impulse is an integral of force with respect to time and work is an integral of force with respect to coordinate. Impulse is a vector, work is a scalar. The SI unit of impulse is kg · m/s; the SI unit of work is kg · m^2/s^2.

☐ In terms of the average force $\overline{\mathbf{F}}$ that acts during a collision, the impulse is given by $\mathbf{J} = \overline{\mathbf{F}}\Delta t$, where Δt is the duration of the collision. This expression is often used to calculate the average force, which is useful as an estimate of the strength of the interaction.

☐ Because Newton's second law is valid, the total impulse acting on an object gives the change in the momentum of the object: $\mathbf{J} = \mathbf{p}_f - \mathbf{p}_i$, where $\mathbf{p}_i$ is the initial momentum and $\mathbf{p}_f$ is the final momentum. Because Newton's third law is valid the impulse of one object on the other is the negative of the impulse of the second object on the first. If external impulses can be neglected, the total momentum of the two objects is conserved during a collision. This means the velocity of the center of mass of the colliding bodies is constant.

☐ For most collisions, the impulse of one colliding object on the other is usually much greater than any external impulse and we may neglect impulses exerted by the environment of the colliding bodies. We may, for example, neglect the effects of gravity and air resistance during the time a baseball is in contact with a bat.

☐ Consider a stream of objects that collide with another object. Examples are bullets hitting a target or the molecules of a gas hitting the walls of a container. Suppose each of the "bullets" has mass m and each suffers the same change in velocity $\Delta\mathbf{v}$. If n "bullets" hit in time Δt, then during that interval the change in momentum of the "target" is given by $\Delta\mathbf{P} = -nm\,\Delta\mathbf{v}$ and the average force exerted by the "bullets" on the "target" is given by $\overline{\mathbf{F}} = \Delta\mathbf{P}/\Delta t = -(n/\Delta t)m\,\Delta\mathbf{v}$.

10–3 Elastic Collisions in One Dimension

☐ The total kinetic energy of the colliding objects may or may not be conserved in a collision. If it is, the collision is said to be **elastic**.

☐ Suppose object 1, with mass m_1 moves with velocity v_{1i} along the x axis and collides elastically with object 2, which has mass m_2 and is initially at rest. After the collision object 1 moves along the x axis with velocity v_{2f} and object 2 moves along the x axis with velocity v_{2f}. The equation that expresses conservation of momentum during the collision is

$$m_1 v_{1i} = m_1 v_{1f} + m_2 v_{2f},$$

and the equation that expresses conservation of kinetic energy during the collision is

$$\tfrac{1}{2} m_1 v_{1i}^2 = \tfrac{1}{2} m_1 v_{1f}^2 + \tfrac{1}{2} m_2 v_{2f}^2.$$

The solution to these two equations is

$$v_{1f} = \frac{m_1 - m_2}{m_1 + m_1} \, v_{1i} \qquad \text{and} \qquad v_{2f} = \frac{2m_1}{m_1 + m_2} \, v_{1i}.$$

☐ Some special cases:

1. If the two masses are the same, then $v_{1f} = 0$ and $v_{2f} = v_{1i}$. The objects have interchanged their velocities.

2. If the target object, object 2, is very massive compared to the incident object, object 1, then $v_{1f} \approx -v_{1i}$ and $v_{2f} \approx (2m_1/m_2)v_{1i}$. The incident object bounces off the target with only a slight loss of speed and the target moves slowly away from the collision.

3. If object 1 is very massive compared to object 2, then $v_{1f} \approx v_{1i}$ and $v_{2f} \approx 2v_{1i}$. The incident object continues with only a slight loss in speed and the target moves away with a speed that is twice the original speed of the incident object.

☐ The velocity of the center of mass of the two-object system is not changed by the collision. If the target object is initially at rest, the velocity of the center of mass Is given by $v_{cm} = m_1 v_{1i}/(m_1 + m_2)$.

☐ If both the target and incident object are initially moving, then

$$v_{1f} = \frac{m_1 - m_2}{m_1 + m_2} v_{1i} + \frac{2m_2}{m_1 + m_2} v_{2i}$$

and

$$v_{2f} = \frac{2m_1}{m_1 + m_2} v_{1i} + \frac{m_2 - m_1}{m_1 + m_2} v_{2i} \, .$$

10–4 Inelastic Collisions in One Dimension

☐ If the total kinetic energy of the colliding objects is not conserved, the collision is said to be **inelastic**. The kinetic energy might increase, as when an explosion takes place, or it might decrease, as when it is converted to internal energy. An explosion in which an object splits into two or more parts as a result of internal forces can be handled in exactly the same manner as a collision: total momentum is conserved but total kinetic energy increases.

☐ After a **completely inelastic collision** the two bodies stick together and move off with the same velocity. During such a collision the loss in total kinetic energy is as large as conservation of momentum allows. That is, the bodies cannot lose a larger fraction of their original kinetic energy and still retain all the original total momentum.

☐ Consider a completely inelastic collision between two objects moving along the x axis. Before the collision object 1, with mass m_1, moves with velocity v_1 and object 2, with mass m_2, moves with velocity v_2. After the collision both objects move along the x axis with velocity V. Since momentum is conserved during the collision $m_1v_1 + m_2v_2 = (m_1 + m_2)V$, so $V = (m_1v_1 + m_2v_2)/(m_1 + m_2)$. Notice that the final speed of the combination *must* be less than the initial speed of object 1.

☐ Kinetic energy is lost during a completely inelastic collision, transformed to internal energy, radiation, etc. Suppose object 2 is initially at rest. The initial total kinetic energy is $K_i = \frac{1}{2}m_1v_1^2$ and the final total kinetic energy is

$K_f = \frac{1}{2}(m_1 + m_2)V^2$. The fractional energy loss is

$$\frac{K_i - K_f}{K_i} = \frac{m_1 v_1^2 - (m_1 + m_2)V^2}{m_1 v_1^2} = \frac{m_2}{m_1 + m_2},$$

where the expression for V in terms of v_1 was substituted.

10–5   Collisions in Two Dimensions

☐   A two-dimensional elastic collision is diagramed below.

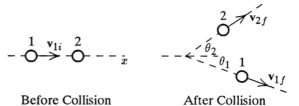

Before Collision   After Collision

Object 2 is initially at rest and object 1 impinges on it with speed v_{1i} along the x axis. Object 1 leaves the collision with speed v_{1f} along a line that is below the x axis and makes the angle θ_1 with that axis. Object 2 leaves the collision with speed v_{2f} along a line that is above the x axis and makes the angle θ_2 with that axis. Take the y axis to be upward in the diagram. Conservation of total momentum **P** leads to the equations $m_1 v_{1i} = m_1 v_{1f} \cos \theta_1 + m_2 v_{2f} \cos \theta_2$ (conservation of P_x) and $0 = -m_1 v_{1f} \sin \theta_1 + m_2 v_{2f} \sin \theta_2$ (conservation of P_y). If the collision is elastic, then $\frac{1}{2}m_1 v_{1i}^2 = \frac{1}{2}m_1 v_{1f}^2 + \frac{1}{2}m_2 v_{2f}^2$. These equations can be solved for three unknowns.

☐   Note that the masses and the initial velocity of object 1 alone do not determine the outcome of a two-dimensional elastic collision. The magnitude and direction of the impulses acting on the object during the collision also play important roles. Often these are not known, so the outcome cannot be predicted. Experimenters usually measure θ_1 or θ_2, then use the equations to calculate other quantities.

☐ If a two-dimensional collision is completely inelastic, the objects stick together after the collision and kinetic energy is not conserved. Suppose that, before the collision object 1 has velocity v_{1i} and object 2 has velocity v_{2i}. After the collision both objects have velocity v_f. Conservation of total linear momentum leads to $m_1 v_{1i} + m_2 v_{2i} = (m_1 + m_2) v_f$. The two corresponding component equations can be solved for two unknowns. Unlike an elastic collision, the masses and initial conditions completely determine the outcome.

10–6 Reactions and Decay Processes

☐ The ideas of this chapter are valid for atomic and subatomic collisions and for spontaneous decays of subatomic particles as well as for collisions of macroscopic objects. In many nuclear and subatomic collision processes, the internal energy and mass energy may change, so they must be included in the conservation of energy equation. The number and type of particles may also change. Changes in particle identities are taken into account in the energy equation by including the energy associated with mass ($E = mc^2$, where c is the speed of light).

NOTES:

Chapter 11
ROTATION

You now begin the study of rotational motion. The pattern is similar to that used for the study of linear motion: first learn how to describe the motion, then learn what changes it. Take special care to understand the definitions of the various angular quantities and the roles they play. Concentrate on Newton's second law for rotation and on the work-kinetic energy theorem for rotation.

Important Concepts

☐ angular position
☐ angular displacement
☐ average angular velocity
☐ (instantaneous) angular velocity
☐ average angular acceleration
☐ (instantaneous) angular acceleration
☐ rotational kinetic energy

☐ rotational inertia
☐ torque
☐ Newton's second law for rotation
☐ work done by a torque
☐ work-kinetic energy theorem for rotation

11–2 The Rotational Variables

☐ If a rigid body undergoes pure rotation about a fixed axis, then the path followed by each point on the body is a circle, centered on the axis of rotation.

☐ The **angular position** of the body is described by giving the angle between a **reference line**, fixed to the body, and a non-rotating coordinate axis. If the body rotates around the z axis, the x axis might be chosen as the fixed axis. The reference line then rotates in the xy plane. Usually the angle θ that gives the angular position is measured counterclockwise from the fixed coordinate axis.

☐ If the body rotates from θ_1 to θ_2, then its **angular displacement** is $\Delta\theta = \theta_2 - \theta_1$. If θ_2 is greater than θ_1, then the angular displacement is positive. According to the convention stated above, the body rotated in the counterclockwise direction. If θ_2 is less than θ_1, then the angular displacement is negative and the body rotated in the clockwise direction.

☐ If the body rotates through more than one revolution, θ continues to increase beyond 2π rad. An angle of 2π rad is not equivalent to 0.

☐ If the body undergoes an angular displacement $\Delta\theta$ in time Δt, its **average angular velocity** during the interval is

$$\overline{\omega} = \frac{\Delta\theta}{\Delta t}.$$

Its **instantaneous angular velocity** at any time is given by the derivative

$$\omega = \frac{d\theta}{dt}.$$

According to the convention given above, a positive value for ω indicates rotation in the counterclockwise direction and a negative value for ω indicates rotation in the clockwise direction. Instantaneous angular velocity is usually called simply angular velocity.

☐ If the angular velocity of the body changes with time, then the body has a non-vanishing **angular acceleration**. If the angular velocity changes from ω_1 to ω_2 in the time interval Δt, then the average angular acceleration in the interval is

$$\overline{\alpha} = \frac{\Delta\omega}{\Delta t},$$

where $\Delta\omega = \omega_2 - \omega_1$. The **instantaneous angular acceleration** at any time t is given by the derivative

$$\alpha = \frac{d\omega}{dt}.$$

Instantaneous angular acceleration is usually called simply angular acceleration.

☐ Common units of angular velocity are deg/s, rad/s, rev/s, and rev/min. Corresponding units of angular acceleration are deg/s^2, rad/s^2, rev/s^2, and rev/min^2.

☐ A positive angular acceleration does not necessarily mean the rotational speed is increasing and a negative angular acceleration does not necessarily mean the rotational speed is decreasing. The rotational speed decreases if ω and α have opposite signs and increases if they have the same sign, no matter what the signs are.

☐ The values of $\Delta\theta$, ω, and α are the same for every point in a rigid body. The angular positions of different points may be different, of course, but when one point rotates through any angle, all points rotate through the same angle. All points rotate through the same angle in the same time and their angular velocities change at the same rate.

11–3   Are Angular Quantities Vectors?

☐ Angular velocity and angular acceleration, but not angular position or displacement, can be considered to be vectors along the axis of rotation. To determine the direction of ω, use the right hand rule: curl the fingers of your right hand around the axis in the direction of rotation. Your thumb then points in the direction of ω.

☐ If the body rotates faster as time goes on, then α and ω are in the same direction. If it rotates more slowly, then α and ω are in opposite directions.

☐ Although directions can be assigned arbitrarily to angular displacements, they do not add as vectors. The result of two successive angular displacements around different axes that are not parallel, for example, depends on the order in which the angular displacements are carried out. Thus, the rules of vector addition are not obeyed.

11–4 Rotation with Constant Angular Acceleration

☐ If a body is rotating around a fixed axis with constant angular acceleration α, then as functions of the time t its angular velocity is given by

$$\omega(t) = \omega_0 + \alpha t$$

and its angular position is given by

$$\theta(t) = \theta_0 + \omega_0 t + \tfrac{1}{2}\alpha t^2 .$$

Here ω_0 is the angular velocity at $t = 0$ and θ_0 is the angular position at $t = 0$. The reference line can usually be placed so $\theta_0 = 0$. These two equations can be solved simultaneously to find values for two of the symbols that appear in them. Another useful equation can be obtained by using one of the equations to eliminate t from the other:

$$\omega^2 - \omega_0^2 = 2\alpha\theta ,$$

where θ_0 was taken to be 0.

☐ Any consistent set of units can be used. Thus, θ in degrees, ω in degrees/s, and α in degrees/s^2 as well as θ in radians, ω in radians/s, and α in radians/s^2 or θ in revolutions, ω in revolutions/s, and α in revolutions/s^2 can be used. Do not mix units, however.

11–5 Relating the Linear and Angular Variables

☐ Each point in a rotating body has a velocity and acceleration and these are related to the angular variables. Consider a point in the body a perpendicular distance r from the axis of rotation. If the body turns through the angle θ (in radians), the point moves a distance $s = \theta r$ along its circular path. The speed of the point is $v = ds/dt = (d\theta/dt)r = \omega r$. The units of ω MUST be rad/s. The tangential component of the acceleration of the point is $a_t = dv/dt = (d\omega/dt)r = \alpha r$, where the units of α MUST be rad/s^2.

☐ Because the point is moving in a circular path its acceleration also has a radial component: $a_r = v^2/r$. In terms of the angular velocity, $a_r = \omega^2 r$. In terms of the components a_r and a_t, the magnitude of the total acceleration is given by $a = \sqrt{a_r^2 + a_t^2}$.

☐ Notice that s, v, and a_t are proportional to r. Compared to a point on the rim of a rotating wheel, for example, a point halfway out travels half the distance, has half the speed, and has half the tangential acceleration.

11–6 Kinetic Energy of Rotation

☐ Suppose a rigid body, rotating about a fixed axis, is made up of N particles. The total kinetic energy is given by $K = \sum \frac{1}{2} m_i v_i^2$, where m_i is the mass of particle i and v_i is its speed. Substitute $v_i = \omega r_i$, where r_i is the distance of particle i from the axis, and obtain $K = \frac{1}{2} \left(\sum m_i r_i^2 \right) \omega^2$. ω MUST be in rad/s.

☐ The sum $\sum m_i r_i^2$ is called the **rotational inertia** of the body and is denoted by I. Thus, the kinetic energy of a rotating body is $K = \frac{1}{2} I \omega^2$.

11–7 Calculating the Rotational Inertia

☐ Rotational inertia is a property of a body that depends on the distribution of mass in the body and on the axis of rotation. A particle far from the axis of rotation contributes more to the rotational inertia than a particle of the same mass closer to the axis.

☐ The sum $\sum m_i r_i^2$ may be difficult to evaluate for a body with more than a few particles. Some bodies, however, can be approximated by a continuous distribution of mass and techniques of the integral calculus can be used to find the rotational inertia. The body is divided into a large number of small regions, each with mass dm. The contribution of each region to the rotational inertia is computed as if the region were a particle; then, the results are summed: $I = \int r^2 \, dm$. Table 11–2 of the text gives the rotational inertias of several bodies.

☐ The defining equation for the rotational inertia is a sum over all particles in the body. If we like, we can consider the body to be composed of two or more parts and calculate the rotational inertia of each part about the same axis, then add the results to obtain the rotational inertia of the complete body.

☐ The defining equation for rotational inertia is used to prove the parallel-axis theorem. Consider two identical bodies, one rotating about an axis through the center of mass and the other rotating about an axis that is parallel to the first axis but is a distance h from the center of mass. The rotational inertia I for the second body is related to the rotational inertia I_{cm} for the first by $I = I_{cm} + Mh^2$, where M is the total mass of the body.

☐ The parallel-axis theorem tells us that of all the places we can position the axis of rotation, the one that leads to the smallest rotational inertia is the one through the center of mass and that the rotational inertia increases as the axis moves away from the center of mass (remaining parallel to its original orientation, of course).

11–8 Torque

☐ A **torque** is associated with any force that is applied to an object and is not along the line from the axis of rotation to the point of application. If the force is perpendicular to the rotation axis, then the torque is given by $\tau = rF_t$, where r is the distance from the axis of rotation to the point of application and F_t is the component of the force perpendicular to the line from the axis to the application point. In the following diagram, the torque associated with $\mathbf{F}$ is $\tau = rF \sin \phi$.

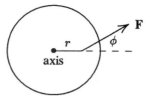

Notice that the magnitude of a torque depends on the distance from the axis to the point of application. A force produces a greater torque if it is applied far from the axis than if it applied closer.

☐ For rotation about a fixed axis, a torque is taken to be positive if it tends to turn the object counterclockwise and negative if it tends to turn the body clockwise. This convention is consistent with the one introduced earlier for the signs of the angular velocity and acceleration.

11–9 Newton's Second Law for Rotation

☐ This law relates the net torque $\sum \tau$ acting on a rigid body to the angular acceleration α of the body:

$$\sum \tau = I\alpha,$$

where I is the rotational inertia of the body about the axis of rotation. It is as important to the study of rotational motion about a fixed axis as $\sum F = ma$ is to the study of one-dimensional translational motion.

☐ Be careful that you use the total torque in this equation. You must identify and sum all the torques that are acting and you must be careful about signs. You must also be careful to use radian measure for the angular acceleration α.

11–10 Work and Rotational Kinetic Energy

☐ Consider a particle traveling counterclockwise in a circular orbit, subjected to a force with tangential component F_t.

As it moves through an infinitesimal arc length ds the magnitude of the work done by the force is $dW = F_t\, ds$. Since $ds = r\, d\theta$, where $d\theta$ is the infinitesimal angular displacement, and $\tau = F_t r$, the expression for the work becomes $dW = \tau\, d\theta$. As the particle travels from θ_1 to θ_2 the work done by the torque is given by the integral

$$W = \int_{\theta_1}^{\theta_2} \tau\, d\theta.$$

☐ The work done by a torque may be positive or negative. If the torque and angular displacement are in the same direction, both clockwise or both counterclockwise, then the work is positive; if they are in opposite directions, one clockwise and one counterclockwise, then the work is negative. If more than one torque acts, sum the works done by the individual torques to find the total work.

☐ In terms of the torque τ and angular velocity ω, the power supplied by the torque is given by $P = dW/dt = \tau\omega$.

☐ A work-kinetic energy theorem is valid for rotational motion: the total work done by all torques acting on a body equals the change in its rotational kinetic energy.

NOTES:

Chapter 12
ROLLING, TORQUE, AND ANGULAR MOMENTUM

This chapter starts with a discussion of objects that roll; that is, simultaneously translate and rotate. Learn the relationship between the velocity of the center of mass and the angular velocity when an object rolls without slipping. Next, rotational motion is discussed using the concept of angular momentum. Learn the definition for a single particle and learn how to calculate the total angular momentum of a collection of particles. Newton's second law for rotation describes the change in angular momentum brought about by the total external torque applied to a body. Angular momentum is at the heart of one of the great conservation laws of mechanics. Pay attention to the conditions for which this law is valid.

Important Concepts

- ☐ rolling
- ☐ (vector) torque
- ☐ angular momentum
- ☐ Newton's second law in angular form
- ☐ conservation of angular momentum

12–1 Rolling

☐ A **rolling** object, such as a wheel, rotates as its center of mass moves along a line. The center of mass obeys Newton's second law: $\sum \mathbf{F} = M\mathbf{a}_{cm}$, where $\sum \mathbf{F}$ is the total force acting on the object, M is the mass of the object, and $\mathbf{a}_{cm}$ is the acceleration of the center of mass. Rotation about an axis through the center of mass is governed by Newton's second law for rotation: $\sum \tau = I\alpha$, where $\sum \tau$ is the total torque acting on the object, I is the rotational inertia of the object, and α is its angular acceleration.

☐ The two motions are related by the forces that act on the wheel. The total force accelerates the center of mass and the total torque (derived from the forces) produces an angular acceleration. If, in addition to the mass and rotational inertia of the body, you know all the forces acting and their points of application you can calculate the acceleration of the center of mass and the angular acceleration about the center of mass. Don't forget the force of friction that may act at the point of contact between the body and the surface on which it rolls.

☐ The velocity of the point on a rolling body in contact with the surface is given by the vector sum of the velocity due to the motion of the center of mass and the velocity due to rotation: $v = v_{cm} - R\omega$, where the forward direction was taken to be positive. If the wheel slips, then $v \neq 0$ and v_{cm} is different from $R\omega$. If it does not slip, then $v = 0$ and $v_{cm} = R\omega$. Furthermore, if the wheel accelerates without slipping, the magnitude of the acceleration of the center of mass a_{cm} is related to the magnitude of the angular acceleration around the center of mass by $a_{cm} = \alpha R$.

☐ A rolling wheel, slipping or not, has both translational and rotational kinetic energy. If the center of mass has velocity v_{cm} and the wheel is rotating with angular velocity ω about an axis through the center of mass, then the total kinetic energy is given by $K = \frac{1}{2}Mv_{cm}^2 + \frac{1}{2}I\omega^2$. If the wheel is not slipping, then $v_{cm} = \omega R$ can be used to write both parts of the kinetic energy in terms of v_{cm} or in terms of ω.

☐ An object that is rolling without slipping on a surface may be considered to be in pure rotation about an axis through the point of contact with the surface. Consider a wheel of radius R, rolling on a horizontal surface. If its center of mass is at its center and its rotational inertia about an axis through that point is I_{cm}, then, according to the parallel axis theorem, its rotational inertia about the point of contact is given by $I = I_{cm} + MR^2$. The total kinetic energy of the wheel is now just its rotational kinetic energy $\frac{1}{2}I\omega^2$. Substitute $\omega = v_{cm}/R$ and the expression for I to obtain $K = \frac{1}{2}Mv_{cm}^2 + \frac{1}{2}I\omega^2$.

12–2   The Yo-Yo

□ A yo-yo is another example of combined rotation and translation. The total force acting (the vector sum of the force of gravity and the force of the string), when entered into Newton's second law for linear motion, gives the acceleration of the center of mass. The total torque acting (due only to the string), when entered into Newton's second law for rotation, gives the angular acceleration about the center of mass. The second law equations, augmented by the condition that the string does not slip ($a = R_0\alpha$, where R_0 is the radius of the axle), can be solved for the acceleration of the center of mass and the angular acceleration about the center of mass.

12–3   Torque Revisited

□ When a force $\mathbf{F}$ is applied to an object at a point with position vector $\mathbf{r}$, relative to some origin, then the **torque** $\boldsymbol{\tau}$ associated with the force is given by the vector product $\boldsymbol{\tau} = \mathbf{r} \times \mathbf{F}$. The magnitude of the torque is given by $\tau = rF\sin\phi$, where ϕ is the angle between $\mathbf{F}$ and $\mathbf{r}$ when they are drawn with their tails at the same point. A right-hand rule gives the direction of the torque: curl the fingers of the right hand so they rotate $\mathbf{r}$ into $\mathbf{F}$ about their tails, through the smallest angle between them. The thumb then points in the direction of the torque.

□ The torque acting on an object is different for different choices of the origin.

□ For an object rotating around a fixed axis, only the component of the torque along the axis is important. For a force that lies in a plane perpendicular to the axis, this component is given by RF_t, where R is the distance from the axis of rotation to the point of application of the force and F_t is the tangential component of the force.

12–4   Angular Momentum

☐ If a particle has momentum $\mathbf{p}$ and position vector $\mathbf{r}$ relative to some origin, then its **angular momentum** is defined by the vector product $\boldsymbol{\ell} = \mathbf{r} \times \mathbf{p}$. Notice that the angular momentum depends on the choice of origin. In particular, if the origin is picked so $\mathbf{r}$ and $\mathbf{p}$ are parallel at some instant of time, then $\boldsymbol{\ell} = 0$ at that instant.

☐ If a particle of mass m is traveling around a circle of radius R with speed v and the origin is placed at the center of the circle, then the magnitude of the angular momentum is $\ell = mRv$. Curl the fingers of the right hand around the axis in the direction the particle is traveling. Then, the thumb points along the axis in the direction of the angular momentum.

☐ For rotation about a fixed axis, the component of the angular momentum along the axis of rotation is the only component of interest. If the orbit of the particle is parallel to the xy plane and the center is on the z axis but is not at the origin, then the angular momentum is not along the z axis. Nevertheless, the z component is $\ell_z = mRv$.

☐ A particle moving along a straight line that does not pass through the origin has a non-zero angular momentum. Its magnitude is given by $\ell = mvd$, where m is the mass of the particle, v is its speed, and d is the distance from the origin to the line of motion.

12–5   Newton's Second Law in Angular Form

☐ If a net torque acts on a particle, then its angular momentum changes with time. Recall that, in terms of the momentum of a particle, Newton's second law can be written $\mathbf{F} = d\mathbf{p}/dt$, where $\mathbf{F}$ is the net force acting on the particle. Take the vector product of this equation with $\mathbf{r}$ to obtain

$\mathbf{r} \times \mathbf{F} = \mathbf{r} \times d\mathbf{p}/dt$. Note that $\mathbf{r} \times \mathbf{F} = \boldsymbol{\tau}$, the net torque acting on the particle. With a small amount of mathematical manipulation you can show that $\mathbf{r} \times (d\mathbf{p}/dt) = d\boldsymbol{\ell}/dt$. Thus, in terms of torque and angular momentum, Newton's second law becomes $\boldsymbol{\tau} = d\boldsymbol{\ell}/dt$.

12–6   The Angular Momentum of a System of Particles

☐   The total angular momentum $\mathbf{L}$ of a system of particles is the vector sum of the individual angular momenta of all the particles in the system: $\mathbf{L} = \sum \boldsymbol{\ell}_i$, where $\boldsymbol{\ell}_i$ is the angular momentum of particle i.

☐   For a system of particles, the time rate of change of the total angular momentum is $d\mathbf{L}/dt = \sum d\boldsymbol{\ell}_i/dt$. Now, the rate of change of the angular momentum of each particle equals the net torque acting on the particle. Thus, the rate of change of the total angular momentum of the system is the vector sum of all torques acting on all particles: $d\mathbf{L}/dt = \sum \boldsymbol{\tau}$. The sum includes the torque exerted by any particle in the system on any other particle in the system as well as external torques, exerted by objects in the environment. Since the torques exerted by two bodies on each other have the same magnitude and opposite directions, internal torques cancel from the sum, leaving only external torques. This means

$$\sum \boldsymbol{\tau}_{\text{ext}} = \frac{d\mathbf{L}}{dt} \, .$$

This is the fundamental equation for a system of particles. Notice that only an external torque can change the total angular momentum of a system. Torques exerted by particles of the system on each other might change the individual angular momenta but not the total.

12–7 The Angular Momentum of a Rigid Body Rotating About a Fixed Axis

☐ For an object rotating with angular velocity ω about the z axis, the z component of the angular momentum, in terms of ω, the mass m_i, and distance $r_{\perp i}$ of each particle from the axis, is the sum $L_z = \sum m_i r_{\perp i} v_i = \sum m_i r_{\perp i}^2 \omega$. In terms of the rotational inertia I of the object, it is $L_z = I\omega$.

☐ Since $L_z = I\omega$ for a body rotating about the z axis, Newton's second law for fixed axis rotation becomes $\tau_{\text{ext } z} = d(I\omega)/dt$. In some cases, the body is rigid and I does not change with time. Then, Newton's second law for fixed axis rotation becomes $\tau_{\text{ext } z} = I\alpha$, where α is the angular acceleration of the body. In this chapter, you will encounter some situations for which I does change.

12–8 Conservation of Angular Momentum

☐ If the total external torque acting on a body is zero, then the change over any time interval of its angular momentum is zero. Angular momentum is then said to be conserved. This is a vector law. One component of the angular momentum may be conserved while other components are not.

☐ If the z component of the total external torque acting on a body rotating about the z axis vanishes, then $I\omega$ is a constant. Consider an ice skater, initially rotating on the points of her skates with her arms extended. By dropping her arms to her sides she decreases her rotational inertia and her angular velocity must increase for her angular momentum to remain the same.

☐ This section concludes with some examples of problems that can be solved using the principle of angular momentum conservation. When attempting to use the conservation principle, you must first check to be sure the total external torque vanishes.

☐ If the rotational inertia of the body changes, write $I_i\omega_i$ for the angular momentum in terms of the initial rotational inertia and angular velocity and $I_f\omega_f$ for the angular momentum in terms of the final rotational inertia and angular velocity. If no external torques act, these are equal. Solve $I_i\omega_i = I_f\omega_f$ for an unknown. See "The Spinning Volunteer" in the text.

☐ If two objects exert torques on each other and no external torques act, write the total angular momentum in terms of the initial angular velocities ($I_1\omega_{1i} + I_2\omega_{2f}$), then write it again in terms of the final angular velocities ($I_1\omega_{1f} + I_2\omega_{2f}$). Equate the two expressions and solve for the unknown. For the spacecraft and flywheel example in the text, the total angular momentum is zero, so $I_{sc}\omega_{sc} + I_{fw}\omega_{fw} = 0$.

☐ In some problems, one object is initially moving along a straight line, as when a child runs and jumps on a merry-go-round. Don't forget to include the angular momentum of this object.

12–9 Quantized Angular Momentum

☐ Angular momentum can have only certain discrete values. For a spinning object, the allowed values are given by $\ell = sh/2\pi$, where h is the Planck constant (6.63×10^{-34} J · s) and s is a half-integer, integer, or zero. The values are so close together that for a macroscopic body they cannot be resolved from each other. For atomic particles, such as protons and electrons, the discrete values can be measured.

☐ The total angular momentum of an isolated system consisting of colliding or decaying fundamental particles is conserved, even if the particles change identity. Conservation of angular momentum often influences the outcome of a collision or decay event.

NOTES:

Chapter 12: Rolling, Torque, and Angular Momentum

Chapter 13
EQUILIBRIUM AND ELASTICITY

This chapter consists largely of applications of Newton's second laws for center of mass motion and for rotation, specialized to situations in which the total force and total torque both vanish. However, several new ideas are introduced to deal with the torque exerted by gravity and with elastic forces: the center of gravity and the elastic moduli. Pay careful attention to what they are and how they are used.

Important Concepts

☐ equilibrium   ☐ compression
☐ static equilibrium   ☐ tension
☐ center of gravity   ☐ Young's modulus
☐ stress   ☐ shear
☐ strain   ☐ shear modulus
☐ yield strength   ☐ hydraulic compression
☐ ultimate strength   ☐ bulk modulus

13–1   Equilibrium

☐ An object is said to be in **equilibrium** if its total momentum **P** and its total angular momentum **L** are both constant. The object is in **static equilibrium** if both these constants are zero. Then, the center of mass is at rest and the object is not rotating.

13–2   The Requirements of Equilibrium

☐ Since the total momentum of an object in equilibrium is constant, the total external force acting on it is zero; since its total angular momentum is constant, the total external torque acting on it is zero. The requirements of equilibrium are

$$\sum \tau_{\text{ext}} = 0 \qquad \text{and} \qquad \sum \mathbf{F}_{\text{ext}} = 0 \,.$$

☐ In applications, these equations are used to determine the external forces and torques on static objects. In some cases, you will be interested in the force or torque exerted by one part of an object on another part. In that event, consider the system to be the part of the object on which the force or torque acts and treat the force or torque as external.

13–3 The Center of Gravity

☐ The force of gravity is often one of the forces acting on an object. Although each particle of the object experiences a gravitational force, for purposes of determining equilibrium these forces can be replaced by a single force acting at a point called the **center of gravity**. If the acceleration **g** due to gravity is uniform throughout the object, the magnitude of the replacement force is Mg, where M is the mass of the object, and the direction of the replacement force is downward, toward the Earth. In addition, the point where the replacement force is applied then coincides with the center of mass. In this case, the position of the center of gravity does not depend on the orientation of the object.

☐ Even if the acceleration due to gravity is not uniform over the object, the sum of the gravitational forces can still be replaced by a single force applied at the center of gravity. Then, however, the center of gravity does not coincide with the center of mass. Furthermore, the position of the center of gravity depends on the orientation of the object. Practically speaking, it is not as useful a concept then.

13–4 Some Examples of Static Equilibrium

☐ In nearly all the examples and problems in this chapter, the forces considered are in the same plane. If you pick an origin in this plane, every torque is perpendicular to the plane and the number of equations to be solved is reduced to three. If the plane of the forces is the xy plane, then the three equations are

$$\sum F_x = 0, \qquad \sum F_y = 0, \qquad \text{and} \qquad \sum \tau_z = 0.$$

They can be solved for three unknowns. The subscripts indicating that the forces and torques are external have been dropped. Usually the last equation is written using τ to stand for τ_z.

☐ Recall that the value of a torque depends on the origin used to specify the point of application of the force. When an object is in equilibrium, however, the torques sum to zero no matter which point is used as the origin. The choice is a matter of convenience, not necessity, but the same origin must be used to compute all torques. If an origin is not specified by the problem statement, place it in the plane of the forces.

☐ Here are the steps you should use to solve problems. First select the object to be considered. Draw a force diagram using arrows to represent all external forces acting on the object. Don't forget the force of gravity and the normal and frictional forces exerted by any objects in contact with the selected object. Use an algebraic symbol to designate the magnitude of each force, known or unknown. Draw each force arrow in the correct direction, if you know it. If you know the line of a force but not its direction, draw the arrow in either direction. If, after you solve the problem, the value turns out to be positive, then you picked the correct direction; if it turns out to be negative, the force is actually opposite the direction you picked. In other cases, you may know only that the force acts in the xy plane. Pick an arbitrary direction, not along one of the coordinate axes. You will then solve for both components.

Because you will need to compute torques, be sure you indicate the point of application of each force by placing the tail of the arrow at the appropriate place on the diagram. Gravity acts at the center of gravity, which for the situations considered here is the same as the center of mass. A string or rod acts at its point of attachment. In many cases, an object makes contact only at a point and that is the point of application of the force it exerts.

Choose a coordinate system for calculating force components. The algebra can usually be simplified by selecting one of the coordinate axes to be in the direction of a force, if any are known. Write $\sum F_x = 0$ and $\sum F_y = 0$ in terms of the individual forces.

Choose an origin for calculating torques. The choice is completely arbitrary but you can usually reduce the algebra considerably by choosing it so that one or more torques vanish. Write $\sum \tau = 0$ in terms of the forces. Pay special attention to the signs of the torques.

Identify the quantities that are known and those that are unknown, then solve the three equations simultaneously for the unknowns.

13–5 Indeterminate Structures

☐ In many cases, the vanishing of the total force and torque is not sufficient to determine the forces acting on an object. From a mathematical viewpoint there are too many unknowns or not enough equations. The additional information that is needed is provided by the equations of elasticity.

13–6 Elasticity

☐ Contact forces, like the normal force of a surface on an object, are **elastic** in nature. That is, they arise from slight deformations of the objects in contact. In some cases, the relationships between elastic forces and deformations are required to understand equilibrium.

☐ A special terminology is used in discussions of elasticity: a force per unit area is called a **stress** and a fractional deformation is called a **strain**.

☐ There are two regimes of deformation. If the stress is less than the **yield strength**, then when the stress is removed, the object returns to it original unstrained condition. If the stress is greater than the yield strength but less than the **ultimate strength**, then when the stress is removed, the object remains deformed. If the stress is greater than the ultimate strength, then the object breaks.

☐ For stresses well below the yield strength, the strain is proportional to the stress and the material is said to be elastic. Elastic materials behave like springs.

☐ Suppose forces that are equal in magnitude and opposite in direction are applied to the opposite ends of a rod. If the forces are perpendicular to the rod faces, the rod will be elongated or compressed, depending on the directions of the forces. Suppose the forces are pulling outward and let ΔL be the elongation of the rod. Also suppose the magnitude of each force is F, the area of each face is A, and the undeformed rod length is L. Then, the stress is given by F/A and the strain is given by $\Delta L/L$. If the object remains elastic (small stress), then the stress is proportional to the strain. The proportionality is written

$$\frac{F}{A} = E\,\frac{\Delta L}{L},$$

where the modulus of elasticity E is called **Young's modulus.** E is a property of the material and does not depend on F, A, L, or ΔL.

☐ Values of E for some materials are listed in Table 13–1 of the text. A large value of E means a large force is required to compress or elongate a given length of sample by a given amount. If E is small and the length and area are the same, only a small force is required.

☐ If the forces applied to the ends of the rod are opposite in direction but parallel to the rod faces, then shear occurs. Suppose one face moves a distance Δx relative to the other, along a line that is parallel to the faces. Also suppose the magnitude of each force is F, the area of each face is A, and the rod length is L. The stress is given by F/A and the strain is given by $\Delta x/L$. Stress is proportional to strain and the proportionality is written

$$\frac{F}{A} = G\,\frac{\Delta x}{L}\,.$$

The modulus of elasticity G is called the **shear modulus**. G is a property of the material. Materials with small shear moduli shear more easily than materials with large shear moduli, provided the lengths and areas of the samples are the same.

☐   If the object is placed in a fluid and pressure is applied uniformly on its surface, then **hydraulic compression** occurs. In this case, the stress is the pressure p in the fluid. This is the same as the force per unit area acting on the object. The strain is $\Delta V/V$, where V is the original volume and ΔV is the change in volume. The modulus of elasticity is called the **bulk modulus** and is denoted by B. The stress-strain relationship is

$$p = B\,\frac{\Delta V}{V}\,.$$

☐   Static equilibrium problems involving deformations are analyzed using the conditions of equilibrium and the appropriate stress-strain relationship.

NOTES:

Chapter 14
GRAVITATION

You will learn about the force law that describes the gravitational attraction of two particles for each other. By considering the attraction of every particle in one object for every particle in a second object the law is extended to deal with objects containing many particles. Then, it is applied to the motion of planets and satellites. You will bring to bear some of the concepts discussed in earlier chapters, chiefly Newton's second law and the principles of energy and momentum conservation.

Important Concepts

☐ Newton's law of gravitation   ☐ escape speed

☐ shell theorems   ☐ Kepler's laws

☐ gravitational potential energy   ☐ principle of equivalence

---

14–1   The World and the Gravitational Force

☐   Gravitational forces are important for the motions of large objects. They hold us and other objects close to Earth's surface, hold planets and satellites in their orbits, and are responsible for the motions of stars and galaxies. At the atomic and subatomic levels, however, they have essentially no influence.

14–2   Newton's Law of Gravitation

☐   The law is: Two particles with masses m_1 and m_2, separated by a distance r, each attract the other with a gravitational force of magnitude

$$F = G\,\frac{m_1 m_2}{r^2}\,,$$

where the universal gravitational constant G has the value $G = 6.67 \times 10^{-11}$ N · m^2/kg^2. The force is proportional the product of the masses and inversely proportional to the square of the particle separation. The force that particle 1 exerts on particle 2 is toward particle 1, along the line that joins the particles; the force that particle 2 exerts on particle 1 is toward particle 2, along the same line. The two forces obey Newton's third law: they have the same magnitude but are in opposite directions.

☐ If the acceleration due to gravity is very nearly uniform throughout a body, then for purposes of studying the motion of the body, we may treat it as a particle located at the center of mass. Thus, small objects in Earth's gravitational field and planets in the Sun's gravitational field are discussed in the text as if they were particles.

☐ The following **shell theorem** follows directly from Newton's law of gravitation: A uniform spherical shell of matter attracts a particle that is *outside* the shell as if all the shell's mass were concentrated at its center. If a particle of mass m is outside a uniform spherical shell of mass M, a distance r from the center of the shell, then the magnitude of the force of the shell on the particle is given by $F = GmM/r^2$.

14–3 Gravitation and the Principle of Superposition

☐ When more than two particles are present, the force on any one of them is the *vector* sum of the individual forces the other particles exert on it. The gravitational force that a continuous distribution of mass exerts on a particle is computed as the integral $\mathbf{F} = \int d\mathbf{F}$, where $d\mathbf{F}$ is the force exerted by an infinitesimal volume element of the distribution. The integral represents a vector sum.

14–4 Gravitation Near Earth's Surface

☐ Earth is nearly a sphere with a spherically symmetric mass distribution, so the gravitational force it exerts on a small object of mass m, outside its surface, a distance r from its center, is very nearly given by $F = GMm/r^2$, where M is the mass of Earth.

☐ If the force of gravity exerted by Earth is the only force acting on a particle of mass m above its surface, Newton's second law becomes $GMm/r^2 = ma_g$, where a_g is the acceleration due to gravity. Thus, $a_g = GM/r^2$. This expression is approximate because Earth's mass distribution is not strictly spherically symmetric.

☐ Even if deviations from a spherical mass distribution can be ignored, the acceleration due to gravity is not the same as the free-fall acceleration g because Earth is spinning. At Earth's surface the difference between the two is $a_g - g = \omega^2 R$, where ω is the angular velocity of Earth and R is the radius of the circle traversed by the object as it rotates with Earth. At the equator R is the radius of Earth; at other latitudes R is less. The weight of the object, as read by a spring balance, is mg.

14–5 Gravitation Inside Earth

☐ A second **shell theorem** states: The gravitational force of a uniform spherical shell of matter on a particle *inside* the shell is zero. The particle can be anywhere inside the spherical hole, not necessarily at its center. Each small region of the shell exerts a gravitational force on the particle but the forces sum to zero.

☐ The two shell theorems can be used to find an expression for the gravitational force on a particle located *within* a spherically symmetric mass distribution. If the particle has mass m and is a distance r from the center, then the magnitude of the force on it is given by $F = GM'm/r^2$, where M' is the mass contained within the sphere of radius r, *not* the total mass of the distribution. The mass outside the sphere of radius r does not exert a net force on the particle.

☐ If the mass distribution is uniform and has mass density ρ, then $M' = (4\pi/3)\rho r^3$. If the distribution has radius R and mass M, then $M' = (r^3/R^3)M$.

14–6 Gravitational Potential Energy

☐ The force of gravity is a conservative force, so a potential energy is associated with a gravitational interaction. If two point masses m and M are brought from infinite separation to a separation r, the work done by the gravitational forces they exert on each other is $W = GMm/r$ and their gravitational potential energy at the final configuration is given by

$$U(r) = -\frac{GMm}{r},$$

where U was taken to be zero at infinite separation. This energy is associated with the *pair* of masses, NOT with either mass alone.

☐ Potential energy is a scalar. The total gravitational potential energy of a system of particles is the sum of the potential energies of each *pair* of particles in the system. This is the work done by an external agent to assemble the particles from infinite separation, starting and ending with the particles at rest. The external agent must do negative work since the mass already in place attracts any new mass being brought in and the agent must pull back on it.

☐ If the particles of a system interact only via gravitational forces and no external forces act, then the total mechanical energy of the system is conserved. As the particles move, the sum of their kinetic energies and the total gravitational potential energy remains the same.

☐ For a two-particle system, the total mechanical energy consists of three terms, corresponding to the kinetic energy of each particle and the potential energy of their interaction. Suppose that at some instant particle 1 (with mass m_1) has speed v_1, particle 2 (with mass m_2) has speed v_2, and the particles are a distance r apart. Then, the mechanical energy of the system is given by

$$E = \tfrac{1}{2}m_1 v_1^2 + \tfrac{1}{2}m_2 v_2^2 - \frac{GMm}{r} \,.$$

At another time v_1, v_2, and r may have different values but the sum of the three terms is the same.

☐ The **escape speed** is the minimum initial speed that an object must be given at the surface of Earth (or other large mass) in order to get far away. The initial gravitational potential energy is $-GMm/R$, where M is the mass of Earth, R is its radius, and m is the mass of the object. The final potential energy is zero (the object is far away). The initial kinetic energy is $\tfrac{1}{2}mv^2$, where v is the escape speed, and the final kinetic energy is zero if v is to have its minimum value. Conservation of energy yields $\tfrac{1}{2}mv^2 - GMm/R = 0$, so

$$v = \sqrt{\frac{2GM}{R}} \,.$$

14–7 Planets and Satellites: Kepler's Laws

☐ Motions of the planets are controlled by gravity, due chiefly to the Sun. If we neglect the influence of other planets, the motion of a planet is described by Kepler's three laws:

1. **Law of orbits**: The orbit of a planet is an ellipse, with the Sun at one focus. This law is a direct consequence of the $1/r^2$ nature of the force law.

2. **Law of areas**: The line that joins the planet and the Sun sweeps out equal areas in equal times. This law is a direct consequence of the principle of angular momentum conservation.

3. **Law of periods**: The square of the period is proportional to the cube of the semimajor axis of the orbit.

☐ A planetary orbit can be described by giving its semimajor axis a, which is half the greatest distance across the orbit, and its eccentricity e, which is defined so that ea is half the distance between the foci. The point of closest approach to the Sun is called the **perihelion** and the distance of this point from the Sun is given by $R_p = a(1 - e)$. The point of maximum distance from the Sun is called the **aphelion** and the distance of this point from the Sun is given by $R_a = a(1 + e)$. For circular orbits, $e = 0$ and $R_a = R_p = a$. An eccentricity of nearly 1 corresponds to an ellipse that is much longer than it is wide.

☐ The law of areas can be used to relate a planet's speed at one point to its speed at another point. The simplest relationship holds for points that are the greatest and least distances from the Sun because at these points the velocity is perpendicular to the position vector from the Sun. In terms of the mass m of the planet, its distance r from the Sun, and its speed v, the magnitude of the angular momentum at one of these points is $L = mrv$, where the origin was placed at the Sun. If v_p is the speed at perihelion and v_a is the speed at aphelion, then conservation of angular momentum yields $R_a v_a = R_p v_p$.

☐ The acceleration of a planet in a circular orbit of radius r is given by $a = v^2/r = 4\pi^2 r/T^2$, where $2\pi r/T$ was substituted for v. Here T is the period. The gravitational force is $F = GMm/r^2$, where M is the mass of the Sun and m is the mass of the planet. Newton's second law yields $GMm/r^2 = 4\pi^2 mr/T^2$, so $T^2 = (4\pi^2/GM)r^3$. This equation is valid for elliptical orbits if r is replaced by the semimajor axis a.

☐ Asteroids and recurring comets in orbit around the Sun and satellites (including the Moon) in orbit around Earth or another planet also obey Kepler's laws. When the two bodies have comparable mass, as for example the two stars in a binary star system, each travels in an elliptical orbit around their center of mass.

14–8 Satellites: Orbits and Energy

☐ For a satellite in a circular orbit of radius r around a Sun of mass M, the acceleration is v^2/r and Newton's second law yields $GMm/r^2 = mv^2/r$. Thus, the kinetic energy is $K = \frac{1}{2}mv^2 = GMm/2r$. If the potential energy is zero for infinite separation, then the potential energy for an orbit of radius r is $U = -GMm/r$ and the mechanical energy is

$$E = K + U = \frac{GMm}{2r} - \frac{GMm}{r} = -\frac{GMm}{2r} \ .$$

E is negative, indicating that the planet is bound to the Sun and does not have enough kinetic energy to escape.

☐ If the orbit is not circular, the mechanical energy is given by $E = -GMm/2a$, where a is the semimajor axis of the orbit. The kinetic energy is

$$K = E - U = \left(\frac{GMm}{r} \right) \left(\frac{1}{r} - \frac{1}{2a} \right) \ .$$

As the planet moves, its distance from the Sun varies; both the kinetic and potential energies vary but the sum remains constant.

14–9 Einstein and Gravitation

☐ Einstein hypothesized that the force of gravity is actually a result of the curvature of space (or more accurately, space-time) and that the curvature is brought about by the presence of mass.

☐ A fundamental postulate of his general theory of relativity is known as the **principle of equivalence** and states that it is impossible to distinguish between a gravitational field and an accelerating reference frame. When we see an object accelerate, we cannot tell if we are at rest in an inertial frame and the object is being attracted gravitationally or if there is no gravitational force and we are accelerating relative to an inertial frame.

☐ The principle of equivalence leads to a prediction that the path of light must bend as it passes close to a massive body. This prediction has been confirmed by observation.

NOTES:

Chapter 15
FLUIDS

Here you study gases and liquids, first at rest and then in motion. Learn well the definitions of pressure and density, then pay particular attention to the variation of pressure with depth in a fluid. Use the concepts to understand two of the most basic principles of fluid statics: Archimedes' and Pascal's principles. The continuity and Bernoulli equations are fundamental for understanding fluids in motion. They express the relationship between pressure, velocity, density, and height at points in a moving fluid. To understand these equations, you must first understand the ideas of streamlines and tubes of flow.

Important Concepts

☐ fluid   ☐ streamline

☐ density   ☐ tube of flow

☐ pressure   ☐ mass flow rate

☐ Pascal's principle   ☐ volume flow rate

☐ Archimedes' principle   ☐ equation of continuity

☐ buoyant force   ☐ Bernoulli's equation

☐ steady, incompressible flow

15–2   What is a Fluid?

☐   Both liquids and gases are classified as **fluids**. Both conform to the sides and bottom of any container in which they are placed. Solids, on the other hand, retain their shapes even when they are not in containers.

15–3 Density and Pressure

☐ If a small volume ΔV of fluid has mass Δm, then its **density** ρ at that place is given by $\rho = \Delta m/\Delta V$. The definition should include a limiting process in which the volume shrinks to a point. Thus, density is defined at each point in a fluid and may vary from point to point. If the density is uniform, then $\rho = M/V$, where M is the mass and V is the volume of the entire fluid. Density is a scalar.

☐ The fluid in any region exerts an outward force on the fluid or container wall that bounds the region. The force on any small surface area ΔA is proportional to the area and is perpendicular to the surface. If ΔF is the magnitude of the force on the area ΔA, then the **pressure** p exerted by the fluid at that place is defined by the scalar relationship $p = \Delta F/\Delta A$. Strictly, the pressure is the limit as the area ΔA tends toward zero. Thus, pressure is defined at each point and may vary from point to point.

☐ The SI unit of pressure (N/m^2) is called a pascal and is abbreviated Pa. Other units are: 1 atmosphere (atm) $= 1.013 \times 10^5$ Pa and 1 mm of Hg $= 1$ torr $= 133.3$ Pa.

☐ If a fluid is compressible, its density depends on the pressure. Great pressure is required to change the densities of most liquids. Gases, on the other hand, are readily compressible.

15–4 Fluids at Rest

☐ Pressure varies with depth in a fluid subjected to gravitational forces. Consider an element of fluid at height y in a larger body of fluid. Suppose the upper and lower faces of the element each have area A and the element has thickness Δy. If the fluid has density ρ, the mass of the element is $\rho A \, \Delta y$ and the force of gravity on it is $\rho g A \, \Delta y$. The pressure at the upper face is $p(y + \Delta y)$ so the downward force of the fluid there is $p(y + \Delta y)A$. The pressure at the lower face is $p(y)$ so the upward force of the fluid there is $p(y)A$.

Since the element is in equilibrium the net force must vanish: $p(y + \Delta y)A + \rho g A \, \Delta y - p(y)A = 0$. In the limit as Δy becomes infinitesimal, this equation yields

$$\frac{\mathrm{d}p}{\mathrm{d}y} = -\rho g \, .$$

The negative sign indicates that the pressure is less at points higher in the fluid than at lower points.

☐ If the fluid is incompressible, then the density is the same everywhere and the pressure difference between any two points in the fluid, at heights y_1 and y_2 respectively, is given by

$$p_2 - p_1 = -\rho g(y_2 - y_1) \, .$$

If p_0 is the pressure at the upper surface of an incompressible fluid, then

$$p = p_0 + \rho g h$$

is the pressure a distance h below the surface.

☐ The pressure is same at any points that are at the same height in a homogeneous fluid. If the fluid is inhomogeneous, as it is if it consists of layers of immiscible fluids with different densities, you must apply the expression for $p(y)$ to each layer separately.

☐ Although atmospheric pressure varies with height, the variation is negligible over distances on the order of meters. The upper surface in each arm of an open U-tube is at atmospheric pressure, even if the surfaces are at different heights.

15–5 Measuring Pressure

☐ A mercury barometer consists of a vertical tube with its open end inserted beneath the surface of a mercury pool with its surface exposed to the air. The region of the tube above the mercury column is evacuated and the pressure there is zero. The height h of the mercury column in the tube is measured from the top of the pool. If the atmospheric pressure is p_0, then $p_0 = \rho g h$, where ρ is the density of mercury. The height h is directly proportional to the atmospheric pressure.

☐ An open-tube manometer is essentially a U-tube that is open at both ends and is filled with an incompressible fluid. One end is inserted into the region where the pressure is to be measured while the other end is held vertically. The difference h in the heights of the fluid in the two arms is measured. If the pressure to be measured is p and atmospheric pressure is p_0, then $p - p_0 = \rho g h$, where ρ is the density of the fluid in the manometer.

☐ A mercury barometer measures **absolute pressure** while an open-tube manometer measures **gauge pressure** (the difference between absolute and atmospheric pressure.

15–6 Pascal's Principle

☐ **Pascal's principle** is: If the external pressure applied to the surface of a fluid changes, then the pressure everywhere in the fluid changes by the same amount.

☐ A hydraulic jack is essentially a tube that is narrow at one end and wide at the other, containing an incompressible fluid and fitted with pistons at both ends. When a small force F_i is applied to the piston at the narrow end, the fluid exerts a much larger force F_o on the piston at the wide end. The change in pressure at the narrow end is F_i/A_i, where A_i is the area of the tube there. The change in pressure at the wide end is F_o/A_o, where A_o is the area there. According to Pascal's principle, these changes must be the same, so $F_i/A_i = F_o/A_o$. Since A_o is much greater than A_i, F_o is much greater than F_i. Hydraulic jacks are used to lift heavy objects, such as automobiles, and to operate brake systems on cars.

☐ Suppose the distances moved by pistons of a hydraulic jack are d_i (at the input end) and d_o (at the output end). If the fluid is incompressible, then the volume of fluid does not change and $d_i A_i = d_o A_o$. Multiply the left side of this equation by F_i/A_i and the right side by the equal quantity F_o/A_o to obtain $F_i d_i = F_o d_o$. The work done by the input force is the same as the work done by the output force.

15–7 Archimedes' Principle

☐ **Archimedes' principle** is: A fluid exerts an upward buoyant force on any object that is partially or wholly immersed in it and the magnitude of the force is equal to the weight of the fluid displaced by the body.

☐ The buoyant force on an object is a direct result of the pressure exerted on it by the fluid and depends on the variation of pressure with depth. The pressure at the bottom of the object is greater than the pressure at the top and the net buoyant force is upward. If the submerged portion of an object is replaced by an equal volume of fluid, the fluid would be in equilibrium and the net force associated with the pressure of surrounding fluid would equal the weight of the fluid that replaced the object.

☐ To calculate the buoyant force acting on a given object, first find the submerged volume. If the object is completely surrounded by fluid, this is the volume of the object. If the object is floating on the surface, it is the volume that is beneath the surface. If the submerged volume is V_s, then the magnitude of the buoyant force is

$$F_b = \rho_f g V_s \,,$$

where ρ_f is the density of the fluid.

☐ For purposes of calculating the torque on a floating object, the buoyant force of the fluid on the object can be treated as a single force acting at a point called the **center of buoyancy**. To find this point, replace the submerged portion of the object with fluid and find the center of mass of the replacement fluid. The force of gravity, on the other hand, can be treated as a single force acting at the center of mass of the body. If buoyancy and gravity together produce a net torque, then the object tilts.

15–8   Ideal Fluids in Motion

☐ Fluid flow can be categorized according to whether it is steady or nonsteady, compressible or incompressible, viscous or nonviscous, and rotational or irrotational. This chapter deals chiefly with ideal flow: steady, incompressible, nonviscous, and irrotational.

☐ The velocity and density of a fluid in **steady flow** do not depend on the time. If we follow all the particles that eventually get to any selected point, we find they all have the same velocity as they pass that point, regardless of their velocities when they are elsewhere. In addition, particles flow into and out of any volume in such a way that the mass in the volume at any time is the same as at any other time.

☐ If the flow is **incompressible**, the density does not depend on either time or position. It is the same everywhere in the fluid and retains the same value through time.

☐ If the flow is nonviscous, no part of the fluid exerts resistive forces on neighboring parts and the fluid does not exert a resistive force on an object in it.

☐ If the flow is irrotational, a small test body placed in it does not rotate about an axis through it.

15–9   Streamlines and the Equation of Continuity

☐ A **streamline** is the path traced out by a moving fluid particle. In steady flow, streamlines are fixed curves in space. Every fluid particle on the same streamline passes through the same sequence of points and at each point has the same velocity as other particles when they are at the point. At any point the fluid velocity is tangent to the streamline through the point.

☐ A **tube of flow** is a tube bounded by streamlines. No particles ever cross the boundaries of a tube of flow. Since streamlines do not cross the boundaries of a tube of flow, they crowd closer together in narrow portions of a tube than in wide portions.

☐ If the fluid passing through the cross section of a narrow tube of flow has density ρ and speed v, then the mass of fluid that passes the cross section per unit time is given by $\rho A v$, where A is the cross-sectional area of the tube. The volume of fluid that passes the cross section per unit time is given by $A v$. The former quantity is called the **mass flow rate** and the later is called the **volume flow rate** and is denoted by R.

☐ In steady flow, the mass of fluid in any volume does not change, so the mass flow rate must be the same for every cross section of a tube of flow. Thus, $\rho A v = $ constant along a tube of flow. This is the **equation of continuity** in its most general form. For incompressible flow, the volume flow rate is the same everywhere along a tube of flow and the equation becomes $A v = $ constant along a tube of flow.

☐ The equation of continuity implies that fluid particles have greater speed in a narrow portion of a tube than in a wider portion. Because streamlines crowd closer together in narrow portions of tubes of flow than in wider portions we conclude that a high concentration of streamlines corresponds to a high fluid speed and a low concentration corresponds to a low fluid speed.

15–10 Bernoulli's Equation

☐ **Bernoulli's equation** for ideal flow tells us that the quantity $p + \frac{1}{2}\rho v^2 + \rho g y$ has the same value at every point along a streamline. The value of the quantity may be different for different streamlines.

☐ This equation arises directly from the work-kinetic energy theorem, applied to the fluid in a narrow tube of flow. The term p appears because the force exerted by neighboring portions of fluid do work, the term $\rho g y$ appears because gravity does work if the height of the tube varies, and the term $\frac{1}{2}\rho v^2$ appears because the kinetic energy of the fluid changes if work is done on it.

☐ If the tube does not change height, the gravitational term is not needed and Bernoulli's equation becomes $p + \frac{1}{2}\rho v^2 =$ constant along a streamline. This equation indicates that the pressure is high where the fluid speed is low and is low where the fluid speed is high. Combining this result with the continuity equation, we conclude that for steady, incompressible, horizontal flow, the pressure is low where a tube of flow is narrow and high where it is wide.

NOTES:

Chapter 16
OSCILLATIONS

This chapter is about objects whose motions are repetitive. Pay attention to the meanings of the terms used to describe simple harmonic motion: amplitude, period, frequency, angular frequency, and phase constant; pay attention to the transfer of energy from kinetic to potential and back again as the object moves; and also pay attention to the form of the force law that leads to this type of motion.

Important Concepts

☐ oscillation

☐ simple harmonic motion

☐ amplitude

☐ angular frequency

☐ frequency

☐ period

☐ phase

☐ phase constant

☐ angular oscillator

☐ simple pendulum

☐ physical pendulum

☐ damped oscillations

☐ forced oscillations

☐ resonance

16–2 Simple Harmonic Motion

☐ If an object moves along the x axis in **simple harmonic motion**, its coordinate x as a function of time t is given by

$$x(t) = x_m \cos(\omega t + \phi),$$

where $x_m, \omega,$ and ϕ are constants. A possible function $x(t)$ is graphed below.

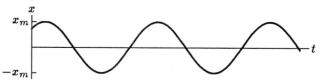

☐ The object moves back and forth between $x = -x_m$ and $x = +x_m$. x_m is called the **amplitude** of the oscillation.

☐ The constant ω is called the **angular frequency** of the oscillation. Since ωt is measured in radians, the unit of ω is rad/s. The **frequency** f of the oscillation gives the number of times the object moves through a complete cycle per unit time. It is measured in hertz ($1\,\mathrm{Hz} = 1\,\mathrm{s}^{-1}$). The frequency and angular frequency are related by $\omega = 2\pi f$. The **period** T of the oscillation is the time for one complete cycle. It is related to the angular frequency and frequency by $T = 1/f$ and $T = 2\pi/\omega$.

☐ The combination $\omega t + \phi$ is called the **phase** of the oscillation and ϕ is called the **phase constant** or **phase angle**. A change in the phase constant simply moves the curve shown on the graph above left or right along the t axis.

☐ An expression for the velocity of the object as a function of time can be found by differentiating the expression for $x(t)$ with respect to time:

$$v(t) = \frac{\mathrm{d}x(t)}{\mathrm{d}t} = -\omega x_m \sin(\omega t + \phi)\,.$$

The maximum speed of the object is given by $v_m = \omega x_m$. The object has maximum speed when its coordinate is $x = 0$. The velocity of the object is zero when its coordinate is $x = -x_m$ and also when its coordinate is $x = +x_m$.

☐ An expression for the acceleration of the object as a function of time can be found by differentiating $v(t)$ with respect to time:

$$a(t) = \frac{\mathrm{d}v(t)}{\mathrm{d}t} = -\omega^2 x_m \cos(\omega t + \phi) = -\omega^2 x(t)\,.$$

The magnitude of the maximum acceleration is $\omega^2 x_m$. The object has maximum acceleration when its coordinate is

$x = -x_m$ and also when its coordinate is $x = +x_m$. This is where the velocity vanishes. The acceleration is zero when the coordinate is $x = 0$. This is where the speed is a maximum.

☐ The amplitude x_m and phase constant ϕ are determined by the initial conditions (at $t = 0$). Since $x(t) = x_m \cos(\omega t + \phi)$ the initial coordinate is given by $x_0 = x_m \cos \phi$ and the initial velocity is given by $v_0 = -\omega x_m \sin \phi$.

x_0 is positive for $-\pi/2 \, \text{rad} < \phi < \pi/2 \, \text{rad}$.

x_0 is negative for $\pi/2 \, \text{rad} < \phi < 3\pi/2 \, \text{rad}$.

v_0 is positive for $\pi \, \text{rad} < \phi < 2\pi \, \text{rad}$.

v_0 is negative for $0 < \phi < \pi \, \text{rad}$.

☐ The equations $x_0 = x_m \cos \phi$ and $v_0 = -\omega x_m \sin \phi$ can be solved for x_m and ϕ. To obtain an expression for x_m, solve the first equation for $\cos \phi$ and the second for $\sin \phi$, then use $\cos^2 \phi + \sin^2 \phi = 1$. To obtain an expression for ϕ, divide the second equation by the first and solve for $\tan \phi$. The results are

$$x_m = \sqrt{x_0^2 + \frac{v_0^2}{\omega^2}} \qquad \text{and} \qquad \tan \phi = -\frac{v_0}{\omega x_0}.$$

☐ Be careful when you evaluate the expression for ϕ. There are always two angles that are the inverse tangent of any quantity, but your calculator only gives the one closest to 0. The other is 180° or π radians away. Always check to be sure $x_m \cos \phi$ gives the correct initial coordinate and $-\omega x_m \sin \phi$ gives the correct initial velocity. If they do not, add π rad to the value you used for ϕ.

16–3   The Force Law for Simple Harmonic Motion

☐ Newton's second law tells us that the force that must be applied to an object of mass m to produce simple harmonic motion is $F(t) = ma(t) = -m\omega^2 x(t)$. The force must be proportional to the displacement from equilibrium and the constant of proportionality must be negative.

☐ Consider a block on the end of spring, moving on a frictionless horizontal surface. If the origin is taken to be at the position of the block when the spring is neither extended nor compressed, the force of the spring on the block is given by $F = -kx$, where k is the spring constant. Since the force is proportional to the displacement and the constant of proportionality is negative we know the motion is simple harmonic. Furthermore, a comparison of $F = -kx$ with $F = -m\omega^2 x$ tells us that the angular frequency of the motion is $\omega = \sqrt{k/m}$. The period is $T = 2\pi/\omega = 2\pi\sqrt{m/k}$.

16–4   Energy in Simple Harmonic Motion

☐ For the block on the end of a spring, an expression for the potential energy as a function of time can be found by substituting $x(t) = x_m \cos(\omega t + \phi)$ into $U = \frac{1}{2}kx^2$:

$$U(t) = \frac{1}{2}kx_m^2 \cos^2(\omega t + \phi).$$

An expression for the kinetic energy as a function of time can be found by substituting $v = -\omega x_m \sin(\omega t + \phi)$ into $K = \frac{1}{2}mv^2$:

$$K(t) = \frac{1}{2}m\omega^2 x_m^2 \sin^2(\omega t + \phi).$$

If $\omega^2 = k/m$ is used, this can also be written

$$K(t) = \frac{1}{2}kx_m^2 \sin^2(\omega t + \phi).$$

☐ Both the potential and kinetic energies vary with time. The potential energy is a maximum when the coordinate is $x = -x_m$ and also when it is $x = +x_m$. Then, the speed is zero and the kinetic energy vanishes. The potential energy is a minimum when $x = 0$. Then, the speed is the greatest and the kinetic energy is a maximum. Notice that the maximum kinetic energy has exactly the same value as the maximum potential energy.

☐ Although both the potential and kinetic energies vary with time, the total mechanical energy $E = K + U$ is constant, as you can see by adding the expressions given above and using $\cos^2(\omega t + \phi) + \sin^2(\omega t + \phi) = 1$. All of the following expressions for the mechanical energy give the same value:

$$E = \tfrac{1}{2}mv^2(t) + \tfrac{1}{2}kx^2(t) = \tfrac{1}{2}kx_m^2 = \tfrac{1}{2}mv_m^2 \,.$$

You will solve some problems by equating two of these expressions and solving for one of the quantities in them.

16–5 An Angular Simple Harmonic Oscillator

☐ An angular simple harmonic oscillator consists of an object suspended by a wire that exerts a torque when it is twisted. The torque is proportional to the angle of twist and the constant of proportionality is negative; it is a restoring torque. If the angular position θ of the object is measured from the its position when the wire is not twisted, then $\tau = -\kappa\theta$, where κ is called the torsion constant of the wire. Newton's second law for rotation becomes $-\kappa\theta = I\alpha$, where I is the rotational inertia of the object. This equation is exactly like the equation for a mass on a spring, except that θ has replaced x, α has replaced a, I has replaced m, and κ has replaced k. The object rotates back and forth in simple harmonic motion with angular frequency $\omega = \sqrt{\kappa/I}$ and period $T = 2\pi/\omega = 2\pi\sqrt{I/\kappa}$.

16–6 Pendulums

☐ A **simple pendulum** consists of a mass m suspended by a string. After the mass is pulled aside and released, it swings back and forth along the arc of a circle with radius equal to the length L of the string. When the string makes the angle θ with the vertical, the tangential component of the gravitational force acting on the mass is $F = -mg\sin\theta$, where the negative sign indicates that the force is pulling the mass toward the $\theta = 0$ position. Newton's second law becomes $-mg \sin \theta = ma$. In terms of the distance s along the arc of the mass from its equilibrium point (the bottom of the arc), $\theta = s/L$ if the angle is measured in radians. Thus, $-g \sin(s/L) = a$, where the mass has been canceled from both sides. The acceleration is not proportional to the displacement and the motion is not strictly simple harmonic. However, if s is much less than L, then $\sin(s/L)$ can be approximated by s/L itself and the equation becomes $-(g/L)s = a$. If s is always small, the motion is very nearly simple harmonic, the angular frequency is $\omega = \sqrt{g/L}$, and the period is $T = 2\pi/\omega = 2\pi\sqrt{L/g}$.

☐ A **physical pendulum** consists of an object that is pivoted about some point other than its center of mass. If h is the distance from the pivot point to the center of mass, the torque acting on the object is $\tau = -mgh \sin \theta$, where θ is the angle between the vertical and the line that joins the pivot and center of mass. Newton's second law for rotation becomes $-mgh \sin \theta = I\alpha$, where I is the rotational inertia of the object. If θ is small, $\sin \theta$ can be approximated by θ itself, in radians. Then, $-mgh\theta = I\alpha$. The angular acceleration is proportional to the angular displacement and the constant of proportionality is negative, so the motion is simple harmonic. The angular frequency is $\omega = \sqrt{mgh/I}$ and the period is $T = 2\pi\sqrt{I/mgh}$.

16–7   Simple Harmonic Motion and Uniform Circular Motion

☐ When a particle moves around a circle with constant speed, the projection of its position vector on the x axis performs simple harmonic motion. Suppose the angular speed of the particle is ω and the radius of the circle is R. Put the origin of the coordinate system at the center of the circle and measure angles counterclockwise from the x axis. If the initial angular position of the particle is ϕ, then the x component of its position vector is given by $x(t) = R\cos(\omega t + \phi)$. You can identify the angular speed ω of the particle with the angular frequency of the oscillation and the radius of the circle with the amplitude of the oscillation.

16–8   Damped Simple Harmonic Motion

☐   Many oscillating systems in nature are damped by a force that is proportional to the velocity. Consider a mass m on the end of spring with spring constant k and subject to the damping force $-bv$, where b is a damping constant and v is the velocity. Newton's second law becomes $-kx - bv = ma$. The solution is

$$x(t) = x_m e^{-bt/2m} \cos(\omega_d t + \phi),$$

where

$$\omega_d = \sqrt{\left(\frac{k}{m}\right)^2 - \frac{b^2}{4m^2}}.$$

The motion is oscillatory but the amplitude $(x_m \, e^{-bt/2m})$ decreases exponentially with time. See Fig. 16–18 of the text for a graph of $x(t)$.

☐   The total mechanical energy of a damped oscillator is not constant. As time goes on, the damping force dissipates energy.

16–9   Forced Oscillations and Resonance

☐   A mass on the end of spring can be driven to oscillate in simple harmonic motion by applying an external force of the form $F_m \cos \omega t$. In this section, the angular frequency of the force is denoted by ω and the natural angular frequency ($\sqrt{k/m}$) of the spring-mass system is denoted by ω_0.

☐   The angular frequency of a forced oscillation is the same as the angular frequency of the driving force and is NOT necessarily the natural angular frequency of the oscillator.

Chapter 16:   Oscillations   **117**

☐ The amplitude of the oscillation depends on the driving frequency, as well as on the driving amplitude F_m. The driving frequency for which the velocity amplitude is a maximum is the same as the natural frequency of the oscillator and is called the **resonance frequency**. It is nearly the same as the frequency for which the amplitude is a maximum. Fig. 16–20 of the text shows how the amplitude depends on the driving frequency.

☐ The amplitude of the motion does not decay with time even though a resistive force is present and energy is being dissipated. The mechanism that drives the oscillator and provides the external force does work on the system and supplies the energy required to keep it going at constant amplitude.

NOTES:

Chapter 17
WAVES — I

In this chapter, you study wave motion, a mechanism by which a disturbance (or distortion) created at one place in a medium propagates to other places. The general ideas are specialized to a mechanical wave on a taut string. Learn what determines the speed of the wave. Also learn about sinusoidal waves, for which the string has the shape of a sine or cosine function. Two or more waves present at the same time and place give rise to what are called interference effects. Special combinations of traveling waves result in standing waves. Learn what these phenomena are and how to analyze them.

Important Concepts

- [ ] wave
- [ ] transverse wave
- [ ] longitudinal wave
- [ ] angular wave number
- [ ] wavelength
- [ ] angular frequency
- [ ] period
- [ ] wave speed
- [ ] power in a wave
- [ ] superposition of waves

- [ ] interference of waves
- [ ] fully constructive interference
- [ ] fully destructive interference
- [ ] intermediate interference
- [ ] phasor
- [ ] standing wave
- [ ] node of a standing wave
- [ ] antinode of a standing wave
- [ ] resonant frequency

17–1 Waves and Particles

- [ ] Energy and momentum are carried from one place to another via two distinct mechanisms: particles and waves. Although particles of the medium move as a mechanical wave passes through, no particle travels with a wave. If a wave is on a taut string, the shape of the string moves from end to end, not any part of the string itself.

17-2 Types of Waves

☐ Wave motion pervades nature. A mechanical wave arises from the motion of particles. Waves on a string and sound waves are examples. An electromagnetic wave arises from disturbances in an electromagnetic field. Radio waves and light are examples. Quantum mechanics deals with matter waves, which are associated with the probability of finding a particle in any region of space.

17-3 Transverse and Longitudinal Waves

☐ Waves are sometimes classified as transverse or longitudinal, according to the direction the medium moves relative to the direction the wave moves. In a **longitudinal** wave, the medium is displaced in a direction that is parallel to the direction the wave travels and in a **transverse** wave it is displaced in a direction that is perpendicular to the direction the wave travels. A wave on a string is transverse; sound waves in air are longitudinal. Many waves, water waves among them, are neither transverse nor longitudinal.

17-4 Wavelength and Frequency

☐ A wave on a string is described by giving the transverse string displacement $y(x,t)$ as a function of position and time, measured from the position of the string when no wave is present. To find the displacement of any point on the string at any time, substitute the value of the coordinate x of the point and the value of the time t into the function.

☐ A sinusoidal wave traveling in the positive x direction has the form $y(x,t) = y_m \sin(kx - \omega t)$, where y_m, k, and ω are constants. Notice the negative sign. The maximum displacement y_m is called the **amplitude** of the wave: every portion of the string moves back and forth between $y = -y_m$ and $y = +y_m$. The constant k is called the **angular wave number** and is related to the shape of the string. The constant ω is the **angular frequency**: every point on the string moves in simple harmonic motion with angular frequency ω. The **frequency** of its motion is given by $f = \omega/2\pi$ and the **period** of its motion is given by $T = 1/f = 2\pi/\omega$.

☐ The graph below shows a sinusoidal wave at time $t = 0$, with mathematical form $y(x,0) = y_m \sin(kx)$. The graph indicates the amplitude y_m and the **wavelength** λ, the distance over which the pattern of the string displacement repeats. The wavelength and angular wave number are related by $k = 2\pi/\lambda$.

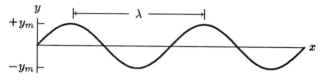

17–5   The Speed of a Traveling Wave

☐ The speed v of a wave is related to the angular wave number k and angular frequency ω by $v = \omega/k$. If the wave is moving in the positive x direction and if the point at x_1 on the string has a given displacement y at time t_1, then at time t_2 the point at $x_2 = x_1 + v(t_2 - t_1)$ will have the same displacement, y. For these two points and times, the phase $kx - \omega t$ has the same value: $kx_1 - \omega t_1 = kx_2 - \omega t_2$, as you can prove by substituting $x_1 + v(t_2 - t_1)$ for x_2 and ω/k for v.

☐ When $2\pi/T$ is substituted for ω and $2\pi/\lambda$ is substituted for k, $v = \omega/k$ becomes $v = \lambda/T$. That is, the wave moves a distance equal to one wavelength in a time equal to one period. The equivalent relationship $v = \lambda f$ is also used.

☐ If the wave is traveling in the negative x direction, the string displacement has the form $y(x,t) = y_m \sin(kx + \omega t)$. Notice the positive sign.

☐ The wave speed is associated with the motion of a distortion in the string shape and is distinct from the velocity of the string itself. The transverse velocity $u(x,t)$ of the point on the string with coordinate x can be found by differentiating $y(x, t)$ with respect to time. If the displacement is given by $y_m \sin(kx - \omega t)$, then $u(x, t) = -\omega y_m \cos(kx - \omega t)$.

☐ Whether the wave is sinusoidal or not, the distortion of the string at $t = 0$ can be described by a function $y = f(x)$. If the wave is traveling with speed v in the positive x direction, then $y(x, t) = f(x - vt)$; if the wave is traveling in the negative x direction, then $y(x, t) = f(x + vt)$. All waves traveling along the x axis are functions of $x - vt$ or $x + vt$, never of x and t in any other combination.

17–6 Wave Speed on a Stretched String

☐ Newton's second law leads directly to an expression for the wave speed in terms of the tension τ in the string and the linear density μ of the string: $v = \sqrt{\tau/\mu}$. The tension in the string is usually determined by the external forces applied at its ends to hold it taut.

☐ The wave speed does NOT depend on the frequency or wavelength. If the frequency is increased (with the tension and linear mass density retaining their values), the wavelength must decrease so that $v = \lambda f$ has the same value.

17–7 Energy and Power of a Traveling String Wave

☐ The mass in a segment of string of infinitesimal length dx is $dm = \mu \, dx$, where μ is the linear mass density. If u is the speed of the segment, then its kinetic energy is $dK = \frac{1}{2} \mu u^2 \, dx$. This energy is transferred to a neighboring segment in time $dt = dx/v$, where v is the wave speed. Thus, the rate with which kinetic energy is carried by the string is $dK/dt = \frac{1}{2} \mu v u^2$. For a sinusoidal wave, with $y(x, t) = y_m \sin(kx - \omega t)$, $dK/dt = \frac{1}{2} \mu v \omega^2 y_m^2 \cos^2(kx - \omega t)$. The average rate over an integer number of periods is given by $\overline{(dK/dt)} = \frac{1}{4} \mu v \omega^2 y_m^2$, since the average value of $\cos^2(kx - \omega t)$ is $1/2$.

☐ The string has potential energy because it stretches as the wave passes by. The average rate with which potential energy is transported is exactly the same as the rate with which kinetic energy is transported so, for a sinusoidal wave, the average rate of energy transport is given by

$$\overline{P} = \frac{1}{2}\mu v \omega^2 y_m^2 \,.$$

It is proportional to the square of the amplitude and to the square of the angular frequency.

17–8 The Principle of Superposition for Waves

☐ When two waves, one with displacement $y_1(x, t)$ and the other with displacement $y_2(x, t)$, are simultaneously on the same string, the displacement of the string is given by their sum: $y(x, t) = y_1(x, t) + y_2(x, t)$, provided the amplitudes are small. This is the **superposition principle**. Displacements, not transmitted powers, add.

☐ A wave of any shape can be constructed as the superposition of sinusoidal waves with different amplitudes and frequencies. This is the basis of Fourier analysis.

17–9 Interference of Waves

☐ The trigonometric identity

$$\sin \alpha + \sin \beta = 2 \sin \left[\tfrac{1}{2}(\alpha + \beta) \right] \cos \left[\tfrac{1}{2}(\alpha - \beta) \right] \,,$$

valid for any angles α and β, is used to derive some results in this and the next chapter.

☐ The superposition of waves leads to **interference** phenomena. Suppose two sinusoidal waves $y_1(x, t) = y_m \sin(kx - \omega t + \phi)$ and $y_2(x, t) = y_m \sin(kx - \omega t)$ are on the same string. The waves are identical except that at every instant the second is shifted along the x axis from the first by an amount that depends on the value of the phase constant ϕ. The trigonometric identity given above can be used to show that the resultant string displacement is

$$y(x, t) = 2y_m \cos(\tfrac{1}{2}\phi) \sin(kx - \omega t + \tfrac{1}{2}\phi).$$

The composite wave is sinusoidal with the same frequency and wavelength as either of the constituent waves.

☐ The amplitude of the resultant wave is $2y_m \cos(\tfrac{1}{2}\phi)$, a function of the phase constant ϕ. The largest possible amplitude occurs if ϕ is 0 or a multiple of 2π rad and its value then is $2y_m$. Maximum amplitude occurs when ϕ is adjusted so the crests of one of the constituent waves fall exactly on the crests of the other. This condition is called **fully constructive interference**.

☐ The smallest possible amplitude occurs when ϕ is an odd multiple of π rad; its value is zero. Minimum amplitude occurs when ϕ is adjusted so the crests of one of the constituent waves fall exactly on the troughs of the other. This condition is called **fully destructive interference**. Other values of ϕ produce **intermediate interference**.

17–10 Phasors

☐ A phasor is a rotating arrow that is used to represent a traveling sinusoidal wave. The length of the arrow, in some scale, is taken to be the amplitude of a sinusoidal traveling wave and the angular velocity is taken to be the angular frequency of the wave. At any instant the angle made by the arrow with the horizontal axis is $kx - \omega t + \phi$. The projection of the arrow on an axis through its tail point behaves like the displacement in the wave.

☐ Phasors can be used to sum waves. Two phasors on the diagram below correspond to waves with amplitudes y_{1m} and y_{2m}, respectively. The waves have the same frequency and wavelength but differ in phase by ϕ. The phasor representing the resultant wave is labeled y_m. The law of cosines gives $y_m^2 = y_{1m}^2 + y_{2m}^2 + 2y_{1m}y_{2m} \cos \phi$.

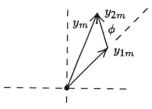

17–11   Standing Waves

☐ In a standing wave, each part of the string oscillates back and forth but the wave does not move. No energy is transmitted from place to place.

☐ A standing wave can be constructed as the superposition of two traveling waves with the same amplitude and frequency, but moving in opposite directions. Let $y_1(x,t) = y_m \sin(kx - \omega t)$ and $y_2(x,t) = y_m \sin(kx + \omega t)$ represent the two traveling waves. The trigonometric identity given above can be used to show that the sum is

$$y = y_1 + y_2 = 2y_m \sin(kx)\cos(\omega t).$$

Each point on the string vibrates in simple harmonic motion with an amplitude that varies with position along the string. In fact, the amplitude of the oscillation of the point with coordinate x is given by $2y_m \sin(kx)$.

☐ At certain points, called **nodes**, the amplitude is zero. Since their displacements are always zero, these points on the string do not vibrate at all. For the standing wave given above, nodes occur at positions for which kx is a multiple of π rad or, what is the same thing, x is a multiple of half a wavelength.

☐ At other points, called **antinodes**, the amplitude is a maximum. Its value is $2y_m$. For the standing wave given above, these points occur at positions for which kx is an odd multiple of $\pi/2$ rad or, what is the same thing, x is an odd multiple of $\lambda/4$.

☐ A standing wave can be generated by reflecting a sinusoidal wave from an end of the string. If such a wave is reflected from a *fixed* end, the reflected and incident waves at that end have opposite signs and cancel each other there. The fixed end is a node of the standing wave pattern. On the other hand, if the same wave is reflected from a *free* end, the incident and reflected waves have the same signs there and the free end of string is an antinode of the standing wave pattern.

17–12 Standing Waves and Resonance

☐ Both ends of a string with both ends fixed are nodes. This means that the traveling waves producing the pattern may have only certain wavelengths, determined by the condition that the length ℓ of the string must be a multiple of $\lambda/2$, where λ is the wavelength. If v is the wave speed for traveling waves on the string, then the standing wave frequencies are $f = v/\lambda = (v/2\ell)n$, where n is a positive integer, called the **harmonic number.**

☐ Standing wave frequencies are called the **resonant frequencies** of the string. If the string is driven at one of its resonant frequencies by an applied sinusoidal force, the amplitude at the antinodes becomes large. The driving force is said to be in resonance with the string. Of course, the string can be driven at other frequencies but then the amplitude remains small.

NOTES:

Chapter 18
WAVES — II

In this chapter, the ideas of wave motion introduced in the last chapter are applied to sound waves. Pay particular attention to the dependence of the wave speed on properties of the medium in which sound is propagating. Completely new concepts include beats and the Doppler shift. Be sure you understand what these phenomena are and how they originate.

Important Concepts

☐ sound wave   ☐ standing sound waves

☐ displacement wave   ☐ beats

☐ pressure wave   ☐ beat angular frequency

☐ sound intensity   ☐ Doppler effect

☐ sound level   ☐ shock wave

18–1 Sound Waves

☐ **Sound waves** are propagating distortions of a medium. In fluids, they are longitudinal: particles of the fluid move back and forth along the line of motion of the wave. In solids, they may be longitudinal, transverse, or neither.

☐ Since particles at slightly different positions move different amounts, the medium becomes compressed or rarefied as a sound wave passes. A sound wave is the propagation of a local increase or decrease in density. Since a change in pressure is associated with a change in density a sound wave is also the propagation of a deviation in local pressure from the ambient pressure.

18–2   The Speed of Sound

☐ The volume of any fluid element changes when the pressure on the element changes. The bulk modulus, a property of the fluid, relates the two changes. If V is the original volume, Δp is the change in pressure, and ΔV is the change in volume, then the bulk modulus of the fluid is $B = -V \, \Delta p / \Delta V$.

☐ The speed of sound in any fluid, including air, is determined by the density ρ and bulk modulus B of the fluid:

$$v = \sqrt{\frac{B}{\rho}} \, .$$

This expression is a direct result of the forces neighboring portions of the fluid exert on each other as the sound wave propagates and can be derived using Newton's second law.

18–3   Traveling Sound Waves

☐ Any of three related quantities can be used to describe a traveling sound wave: the particle displacement s, the deviation of the density from its ambient value $\Delta \rho$, and the deviation of the pressure from its ambient value Δp.

☐ Suppose a sound wave is traveling along the x axis through a fluid and that the displacement of the fluid at coordinate x and time t is given by a known function $s(x, t)$. In the presence of the wave, the fractional change in the volume of any element of the fluid is given by $\Delta V / V = \partial s / \partial x$, the fractional change in the density is given by $\Delta \rho / \rho = -\Delta V / V = -\partial s / \partial x$, and the change in pressure is given by $\Delta p = -B \, \Delta V / V = -B \, \partial s / \partial x$. These expressions can be derived by considering the expansion or contraction of the fluid element in the presence of the wave.

☐ Suppose the displacement is given by $s(x, t) = s_m \cos(kx - \omega t)$, where k is the wave number and ω is the angular frequency. Then, the deviation of the pressure from its ambient value is given by

$$\Delta p(x,t) = -B\,\frac{\partial s}{\partial x} = Bks_m\sin(kx - \omega t)\,,$$

which is sometimes written

$$\Delta p = \Delta p_m\sin(kx - \omega t)\,,$$

where

$$\Delta p_m = Bks_m = v^2\rho ks_m\,.$$

☐ The pressure wave is not in phase with the displacement wave. A fluid element in the neighborhood of a displacement maximum is neither compressed or elongated and the deviation of the pressure from its ambient value is zero there. The elongation or compression is greatest in the neighborhood of a displacement zero and the deviation of the pressure from its ambient value is greatest there.

18–4 Interference

☐ Sound waves obey a superposition principle. Two waves with the same frequency, traveling in the same direction in the same region interfere constructively if their phase difference ϕ has any of the values $2n\pi$ rad, where n is an integer. The interference is completely destructive if ϕ has any of the values $n\pi$ rad, where n is an odd integer.

☐ If the waves are generated by sources that are in phase but they travel to the detector along different paths, they may have different phases at the detector. If one wave travels a distance x to the detector and the other travels a distance $x + \Delta d$, then the phase difference at the detector is given by $\phi = k\,\Delta d$, where k is the angular wave number. Since $k = 2\pi/\lambda$, where λ is the wavelength, $\phi = 2\pi\,\Delta d/\lambda$. Fully constructive interference occurs if Δd is a multiple of λ; fully destructive interference occurs if Δd is an odd multiple of $\lambda/2$.

18–5 Intensity and Sound Level

☐ The **intensity** of a sound wave is the average rate of energy flow per unit cross-sectional area and, for the sinusoidal wave discussed above, is given by

$$I = \tfrac{1}{2}\rho v \omega^2 s_m^2 .$$

This expression can be derived by considering the energy in an infinitesimal element of fluid. Energy is transported with the wave, at the wave speed. The SI units for intensity are W/m^2.

☐ The **sound level** associated with intensity I is defined by $\beta = (10\,\text{dB}) \log(I/I_0)$, where I_0 is the standard reference intensity ($10^{-12}\ W/m^2$), roughly at the threshold of human hearing. Sound level is measured in units of decibels, abbreviated dB.

☐ If $I = I_0$, the sound level is zero. If the intensity is increased by a factor of 10, the sound level increases by 10 dB.

18–6 Sources of Musical Sound

☐ Two sinusoidal traveling sound waves with the same frequency and amplitude but traveling in opposite directions combine to form a standing wave. At a point with coordinate x the particle displacement oscillates with an amplitude that is given by $2s_m \sin(kx)$, where s_m is the amplitude and k is the angular wave number of either of the traveling waves. At a displacement node $kx = 2n\pi$, where n is an integer, and the displacement is always zero. At a displacement antinode $kx = n\pi/2$, where n is an odd integer, and the displacement oscillates between $-2s_m$ and $+2s_m$.

☐ Standing sound waves are created in pipes by the superposition of a sinusoidal wave and its reflection from the end of the pipe. A displacement node exists at a closed end of a pipe. An open end is nearly a displacement antinode.

☐ If both ends of a pipe are open, the wavelengths associated with possible standing waves are such that the pipe length L is a multiple of $\lambda/2$. They are $\lambda = 2L/n$, where n is an integer, and the corresponding standing wave frequencies are

$$f = \frac{v}{\lambda} = \frac{nv}{2L},$$

where v is the speed of sound for the fluid that fills the pipe.

☐ If one end of a pipe is open and the other is closed, the standing wave wavelengths are such that the pipe length is an odd multiple of $\lambda/4$. They are $\lambda = 4L/n$, where n is an odd integer, and the corresponding standing wave frequencies are

$$f = \frac{v}{\lambda} = \frac{nv}{4L},$$

where v is the speed of sound for the fluid that fills the pipe.

☐ A string of a stringed instrument or the air in an organ pipe can vibrate with any superposition of its standing waves. Which combination is present depends on how the vibration is generated. Usually the lowest frequency dominates but higher frequency sound is mixed in. The admixture of higher frequencies gives any instrument the quality of sound peculiar to that instrument and, for example, allows us to distinguish a violin from a piano.

☐ The lowest frequency is called the **fundamental frequency** or **first harmonic**, while higher frequencies are higher harmonics.

18–7 Beats

☐ **Beats** are created by the combination of two sound waves with nearly the same frequency. Let $s_1 = s_m \cos(\omega_1 t)$ represent the displacement at some point due to one of the waves and $s_2 = s_m \cos(\omega_2 t)$ represent the displacement at the same point due to the other wave. The resultant displacement is the sum of the two and, once the trigonometric identity given Chapter 17 is used, it can be written as

$$s(t) = 2s_m \cos(\omega't)\cos(\omega t),$$

where $\omega = (\omega_1 + \omega_2)/2$ and $\omega' = (\omega_1 - \omega_2)/2$. There are two time dependent factors, both periodic. The angular frequency of one is the average of the two constituent angular frequencies. This is the greater of the two angular frequencies and if the two constituent frequencies are nearly the same, it is essentially equal to either of them.

☐ The angular frequency of the other time dependent factor is $\omega' = (\omega_1 - \omega_2)/2$. It depends on the difference in the two constituent frequencies. If ω_1 and ω_2 are nearly the same, this factor is slowly varying. We may think of it as a slow variation in the amplitude of the faster vibration. The effect can be produced, for example, by blowing a note on a horn at the angular frequency ω, but modulating it so it is periodically loud and soft.

☐ A **beat** is a maximum in the *intensity* and occurs each time $\cos(\omega't)$ is $+1$ or -1. Thus, the **beat angular frequency** is $2\omega'$; that is,

$$\omega_{beat} = |\omega_1 - \omega_2|.$$

18–8   The Doppler Effect

☐ Suppose a sustained note with a well-defined frequency f is played by a stationary trumpeter. If you move rapidly *toward* the trumpeter, you will hear a note with a higher frequency. If you move rapidly *away* from the trumpeter, you will hear a note with a lower frequency. Similar effects occur if you are stationary and the trumpeter is moving: the note has a higher frequency if the trumpeter is moving toward you and a lower frequency if the trumpeter is moving away from you. These are examples of the **Doppler effect**.

☐ If a source emits a sound of frequency f and is moving with speed v_S, then the frequency detected by a detector moving with speed v_D is given by

$$f' = f\,\frac{v \pm v_D}{v \mp v_S}\ .$$

You can easily determine which signs to use in any particular situation by remembering that motion of the source toward the detector or the detector toward the source results in detecting a higher frequency while motion of the source away from the detector or the detector away from the source results in hearing a lower frequency than would be heard if both were stationary.

☐ The expression for f' can be derived by considering the number of wave crests that are intercepted by the detector per unit time.

☐ The velocities in the Doppler effect equation are measured relative to the medium in which the wave is propagating (the air, for example). What counts is not the motion of the source relative to the observer but the motions of both the source and observer relative to the medium of propagation. The Doppler effect equation given above is valid only if the motion is along the line joining the source and detector. For motion in other directions, v_D and v_S must be interpreted as components of the velocities along that line.

☐ If a source of sound is moving through a medium faster than the speed of sound in the medium, a **shock wave** is produced. Then, a wavefront has the shape of a cone, with the source at its apex, as shown in Fig. 18–22 of the text. The half angle θ of the cone is given by $\sin\theta = v/v_S$, where v is the speed of sound and v_S is the speed of the source. A shock wave is not produced if $v_S < v$.

18–9   The Doppler Effect for Light

☐ The Doppler effect also occurs for electromagnetic radiation. Light from stars that are moving at high speeds away from the Earth is shifted toward the red and the extent of the shift is used to calculate the speeds of the stars. Doppler shifts of radar waves reflected from moving objects can be used to find their speeds. Police use the effect to detect speeders and TV technicians use it to find the speeds of thrown baseballs.

☐ The change in frequency is not given by the expression derived for the change in frequency of a sound wave, except when the source and detector of a light wave are moving slowly compared with the speed of light. Then,

$$f' = f\left(1 \pm \frac{u}{c}\right) ,$$

where c is the speed of light and u is the relative speed of the source and detector. The same expression, with c replaced by the speed of sound v, is obtained for sound waves if the source and detector are moving slowly compared with the speed of sound.

NOTES:

Chapter 19
TEMPERATURE, HEAT,
AND THE FIRST LAW OF THERMODYNAMICS

You now begin the study of thermodynamics. Be sure you understand how temperature is defined and measured. Pay particular attention to the zeroth law of thermodynamics, which codifies the characteristic of nature that allows temperature to be defined. You will also study the phenomenon of thermal expansion, the familiar change in the dimensions of an object that occurs when its temperature is changed. The first law of thermodynamics, derived from the conservation of energy principle, governs the exchange of energy between a system and its environment. Be sure to distinguish between the two mechanisms of energy transfer, work and heat. Learn how to calculate the heat absorbed or rejected in terms of the heat capacity of the system and learn how to calculate the work done by a system, given the pressure as a function of volume.

Important Concepts

- [ ] thermal equilibrium
- [ ] temperature
- [ ] constant-volume gas thermometer
- [ ] Kelvin temperature scale
- [ ] Celsius temperature scale
- [ ] Fahrenheit temperature scale
- [ ] thermal expansion
- [ ] coefficient of linear expansion
- [ ] coefficient of volume expansion
- [ ] heat
- [ ] heat capacity
- [ ] specific heat
- [ ] molar specific heat
- [ ] heat of fusion
- [ ] heat of vaporization
- [ ] thermodynamic state
- [ ] work done by a gas
- [ ] first law of thermodynamics
- [ ] adiabatic process
- [ ] free expansion
- [ ] heat conduction
- [ ] heat convection
- [ ] heat radiation
- [ ] rate of heat transfer
- [ ] thermal conductivity

19–1 Thermodynamics

☐ Thermodynamics deals chiefly with changes in the internal energies of objects. Temperature is a central concept.

☐ The temperature of any object cannot be lowered beyond a certain limiting value. This value is taken to be zero on the **Kelvin temperature scale**, the temperature scale most widely used in physics. Room temperature on the Kelvin scale is about 300 kelvins (or 300 K).

19–2 The Zeroth Law of Thermodynamics

☐ Temperature is intimately related to the idea of **thermal equilibrium**. If two objects that are not in thermal equilibrium with each other are allowed to exchange energy, they do so and some or all of their macroscopic properties change. When the properties stop changing, there is no longer a net flow of energy and the objects are in thermal equilibrium with each other. They are then at the same temperature.

☐ If two objects are in thermal equilibrium, all their macroscopic properties except temperature might have different values. Suppose the systems are gases. When they are in thermal equilibrium, their pressures may be different, their volumes may be different, their particle numbers may be different, and their internal energies may be different. But their temperatures are the same.

☐ The zeroth law of thermodynamics legitimizes the temperature as a property of a body in thermal equilibrium with another. It is: If bodies A and B are each in thermal equilibrium with a third body C, then they are in thermal equilibrium with each other. Suppose the law were not valid and suppose further that body A and body B are in thermal equilibrium and therefore at the same temperature. If body C is in equilibrium with A but not with B, then the temperature of C is not a well defined quantity and cannot be considered a property of C.

☐ An important consequence of the zeroth law is that it allows us to select some object for use as a thermometer. Suppose that when the thermometer is in thermal equilibrium with body A it is also in thermal equilibrium with body B. Then, because the law is valid, we know that A and B have the same temperature.

19–3 Measuring Temperature

☐ In general, temperature is measured by measuring some property of a system. For an ordinary mercury thermometer, the height of the mercury column is measured. When a constant-volume gas thermometer is used, a bulb containing a gas is put in contact with the object whose temperature is to be measured and the pressure of the gas in the bulb is measured. A reservoir of fluid connected to the barometer is used to ensure the gas in the bulb always has the same volume, no matter what its temperature.

☐ The temperature in kelvins of a constant-volume gas thermometer is taken to be proportional to the pressure: $T = Cp$, where the constant of proportionality C is chosen so $T = 273.16\,\text{K}$ at the triple point of water. Thus, $T = T_3(p/p_3)$, where T_3 is the triple-point temperature and p_3 is the triple-point pressure. Actually the ratio p/p_3 must be evaluated in the limit as the amount of gas in the bulb approaches zero. In this limit, the measured value of T does not depend on the type of gas used.

19–4 The Celsius and Fahrenheit Scales

☐ Three temperature scales are commonly used: Kelvin (or absolute), Celsius, and Fahrenheit. A constant-volume gas thermometer gives the temperature on the Kelvin scale. The Celsius temperature T_C is given in terms of the Kelvin-scale temperature T by $T_C = T - 273.15°$. The Fahrenheit temperature T_F is given in terms of the Celsius temperature by $T_F = \frac{9}{5}T_C + 32°$. A Celsius degree is the same as a kelvin and is 9/5 times as large as a Fahrenheit degree .

19–5 Thermal Expansion

☐ Most solids and liquids expand when the temperature is increased and contract when it is decreased. If the temperature is changed by ΔT, a rod originally of length L changes length by $\Delta L = L\alpha \Delta T$, where α is the **coefficient of linear expansion**. Over small temperature ranges α is essentially independent of temperature.

☐ When the temperature changes by ΔT, the length of every line in an isotropic material changes by the same fraction: $\Delta L/L = \alpha\Delta T$. The fractional change in the length of a scratch on the surface of an isotropic solid is also the same. The fractional change in the diameter D of a hole in an object is given by $\Delta D/D = \alpha \Delta T$.

☐ When the temperature changes by ΔT, the area A of a face of a solid changes by the fraction $\Delta A/A = 2\alpha \Delta T$ and the volume V of the solid changes by the fraction $\Delta V/V = 3\alpha \Delta T$, where α is the coefficient of linear expansion. For a fluid, the fractional change in volume is $\Delta V/V = \beta \Delta T$, where β is the **coefficient of volume expansion**.

☐ Water near 4°C and a few other substances decrease in volume when the temperature is increased. These materials have negative coefficients of thermal expansion in the temperature range for which such a contraction occurs.

19–6 Temperature and Heat

☐ Work and heat are alternate ways of changing the internal energy of a system. **Heat** is the energy that flows between a system and its environment because they are at different temperatures. Thermodynamics deals only with work that changes the internal energy, not with work that changes the motion of a system as a whole.

☐ A sign convention is adopted for work and heat. Work W is taken to be positive when it is done *by* the system and heat Q is taken to be positive when it *enters* the system.

☐ The SI unit of heat is the joule. Other units in common use and their SI equivalents are the calorie ($1\,\mathrm{cal} = 4.186\,\mathrm{J}$), the British thermal unit ($1\,\mathrm{Btu} = 1055\,\mathrm{J}$), and the nutritional calorie ($1\,\mathrm{Cal} = 4186\,\mathrm{J}$).

19–7 The Absorption of Heat by Solids and Liquids

☐ The heat capacity relates the heat transferred into or out of a system to the change in temperature of the system. If during some process heat Q is absorbed and the temperature increases by a small increment ΔT, then the heat capacity of the system for that process is given by $C = Q/\Delta T$.

☐ The heat capacity depends on the kind and amount of material in the system. A related property that depends only on the kind of material is the specific heat, defined by $c = C/m$, where m is the mass of the system. Another is the molar specific heat, defined by $C' = C/n$, where n is the number of moles in the system. A mole is 6.02×10^{23} elementary units (molecules for a gas). This number is called Avogadro's number and is denoted by N_A.

☐ The heat capacity depends on the process by which heat is transferred. It is different for constant volume processes and constant pressure processes, for example.

☐ When two objects, initially at different temperatures, are placed in thermal contact with each other and the composite is isolated from its surroundings, the hotter substance cools, and the cooler substance warms until they reach the same temperature. The heat leaving the hotter object has the same magnitude as the heat entering the cooler object. The algebraic relationship is $Q_A + Q_B = 0$. If object A has mass m_A, specific heat c_A, and initial temperature T_A, then $Q_A = m_A c_A (T_f - T_A)$, where T_f is the final temperature. If object B has mass m_B, specific heat c_B, and initial temperature T_B, then $Q_B = m_B c_B (T_f - T_B)$. The equation $m_A c_A (T_f - T_A) + m_B c_B (T_f - T_B) = 0$ can be solved for the final temperature.

☐ A substance accepts or rejects heat when it changes phase (melts, freezes, vaporizes, or condenses), even though the temperature remains constant during a phase change. The magnitude of the heat accompanying a phase change of a system with mass m is given by $|Q| = mL$, where L is the **heat of transformation.** Q is positive (heat absorbed) for melting and vaporization; it is negative (heat rejected) for freezing and condensing. The **heat of fusion** L_F is the heat per unit mass transferred during freezing and melting. The **heat of vaporization** L_V is the heat per unit mass transferred during boiling or condensing.

19–8   A Closer Look at Heat and Work

☐ The **thermodynamic state** of a gas can be described by giving the values of three thermodynamic variables: pressure, volume, and temperature. The state is changed by doing (positive or negative) work on the gas or by transferring energy as heat between the gas and its environment.

☐ Thermodynamics deals with the work done *by* the system, rather than with the work done *on* the system. These are the negatives of each other. When the volume of a gas changes from V_i to V_f, the work that is done *by* the gas is given by the integral of the pressure p:

$$ W = \int_{V_i}^{V_f} p \, dV . $$

This expression is valid only if the process is carried out so the gas is nearly in thermal equilibrium at all times. Only then is the pressure well defined throughout the process. W is positive if $V_f > V_i$ and is negative if $V_f < V_i$.

☐ The process can be plotted as a curve on a graph with p as the vertical axis and V as the horizontal axis. Then, the work W is the area under the curve.

☐ The work done on the system is different for different processes, represented as different functional dependencies of

p on V and plotted as different curves on a p-V diagram. To calculate the work, you must know how the pressure varies as the volume changes.

19–9 The First Law of Thermodynamics

☐ The **first law of thermodynamics** postulates the existence of an internal energy E_{int} and states that its change as the system goes from any initial thermal equilibrium state to any final thermal equilibrium state is independent of how the change is brought about. It then equates the change in the internal energy to $Q - W$, evaluated for any path between the initial and final states. For an infinitesimal change in state, $dE_{int} = dQ - dW$. Positive work done by the system and heat leaving the system (dW positive and dQ negative) both tend to decrease the internal energy.

☐ When a system undergoes a change from one equilibrium state to another, the work done by the system and the heat absorbed by the system depend on the process and may be different for different processes. Changes in the internal energy, like changes in temperature, pressure, and volume, do not depend on the process.

☐ Because a change in the internal energy is a function of only the initial and final states, the process used to change states is immaterial for its calculation. At intermediate stages of the process the system might not even be in thermal equilibrium.

19–10 Some Special Cases of the First Law of Thermodynamics

☐ If a system undergoes a change of state at constant volume, then the work done by it is zero and the change in internal energy is related to the heat absorbed by $\Delta E_{int} = Q$. When heat Q is absorbed at constant volume, the internal energy increases by exactly that amount.

☐ If the change of state is **adiabatic**, then the heat absorbed is zero and the change in internal energy is related to the work by $\Delta E_{int} = -W$. When work W is done by the system in an adiabatic process, the internal energy decreases by exactly that amount.

☐ If the process is cyclic, so the initial and final states are the same, then the change in internal energy is zero and the work done by the system is related to the heat absorbed by $W = Q$.

☐ The adiabatic free expansion of a gas is a process for which the system is not in thermal equilibrium during intermediate stages. Initially a partition confines the gas to one part of its container, which is thermally insulated. When the partition is removed, the gas expands into the other part. For this process, $W = 0$ and $Q = 0$. According to the first law, $\Delta E_{int} = 0$.

19–11 Heat Transfer Mechanisms

☐ There are three mechanisms by which heat flows from one place to another:

conduction: Energy is passed from atom to atom in an object during atomic collisions. Example: heat is conducted from the hot to the cold end of a metal rod.

convection: Energy is carried from one place to another by the flow of a fluid. Example: air above a hot radiator is warmed and rises, carrying energy toward the ceiling.

radiation: A hot object emits electromagnetic radiation, which is absorbed by a cooler object. Example: the Earth absorbs radiation from the sun.

☐ If heat Q is transferred in time t, then the **rate of heat transfer** H over that time is given by $H = Q/t$.

☐ Consider a homogeneous bar of material of length L, with one end held at temperature T_H (hot) and the other held at temperature T_C (cold). The temperature in the bar varies from point to point along its length. The rate of heat flow through the bar is proportional to the temperature difference of its ends, proportional to its cross-sectional area A, and inversely proportional to its length L. The constant of proportionality k is called the **thermal conductivity** of the material. Mathematically,

$$H = kA \frac{T_H - T_C}{L}.$$

The thermal conductivity is essentially independent of the temperature over small temperature ranges.

☐ Building materials are often characterized by their **thermal resistances** or R values rather than their thermal conductivities. For a slab of material, the R value is related to the thermal conductivity k by $R = L/k$, where L is the thickness of the slab.

NOTES:

Chapter 20
THE KINETIC THEORY OF GASES

Kinetic theory is used to relate macroscopic quantities like pressure, temperature, and internal energy to the energies and momenta of the particles that comprise a gas. Here you will gain an understanding of the relationship between the microscopic and macroscopic descriptions of a gas. The example used throughout this chapter and the next is known as an ideal gas. Pay careful attention to its properties, particularly its molar heat capacities. Also learn about the average distance a molecule in a gas travels between collisions with other molecules and about the distribution of molecular speeds in a gas.

Important Concepts

☐ Avogadro's number

☐ ideal gas

☐ ideal gas law

☐ root-mean-square speed

☐ translational energy
of an ideal gas

☐ mean free path

☐ Maxwell speed distribution

☐ molar heat capacities
of ideal gases

☐ equipartition of energy

☐ adiabatic expansion

20–2 Avogadro's Number

☐ The quantity of matter in a system of identical particles might be given as the number of particles or as the number of moles. The number of molecules in a mole is Avogadro's number ($N_A = 6.02 \times 10^{23}$). If a gas contains N molecules, then it contains N/N_A moles. If the mass of one mole is M, then the mass of one molecule is $m = M/N_A$. A gas containing N molecules has a total mass of Nm and a gas containing n moles has a total mass of nM.

20–3 Ideal Gases

☐ The molecules of an **ideal gas** are point-like, taking up a negligible portion of the volume, and they interact only in elastic collisions of very short duration. The potential energy of their interactions is negligible compared to their kinetic energies. Real gases become more nearly ideal as the number of molecules per unit volume decreases.

☐ The **ideal gas law** (or ideal gas equation of state) expresses the relationship between the absolute temperature T, pressure p, volume V, and number of moles n for an ideal gas:

$$pV = nRT \, .$$

R is the universal gas constant ($8.31 \, \text{J/mol} \cdot \text{K}$).

☐ An **isothermal process** is one during which the temperature remains constant. For an ideal gas, $p = nRT/V$ and the work it does during an isothermal change in volume is

$$W = \int_{V_i}^{V_f} p(V) \, \mathrm{d}V = \int_{V_i}^{V_f} \frac{nRT}{V} \, \mathrm{d}V = nRT \ln \frac{V_f}{V_i} \, .$$

20–4 Pressure, Temperature, and RMS Speed

☐ The pressure in a gas is intimately related to the average of the squares of the speeds of the molecules:

$$p = \frac{nM\overline{v^2}}{3V} \, ,$$

where n is the number of moles, M is the molar mass, V is the volume, and $\overline{v^2}$ is the average value of the squares of the molecular speeds.

☐ This expression for the pressure can be derived by calculating the average force per unit area exerted by the molecules of a gas on the container walls. Assume that at each collision between a molecule and a wall the normal component of the molecule's momentum is reversed. Compute the change in momentum per unit time to find the average force exerted by the molecule on the wall. When this is divided by the area of the wall, the result is the contribution of the molecule to the pressure.

☐ Consider an ideal gas in a cubical container with edge L. If m is the mass of a molecule, v_x is the component of its velocity normal to the wall, and Δt is the time from one collision to the next with the same wall, then the average force exerted by the molecule on the wall is given by $F = 2mv_x/\Delta t$. The time between collisions is $\Delta t = 2L/v_x$, so the average force of the molecule on the wall is $F = (2mv_x)/(2L/v_x) = mv_x^2/L$ and its contribution to the pressure is $F/L^2 = mv_x^2/L^3$. This is now summed over all molecules. Replace the sum of v_x^2 with the product of its average value and the number of molecules nN_A to obtain $p = nmN_A\overline{v_x^2}/L^3$. Finally, replace the average of v_x^2 with one third the average of v^2, mN_A with M, and L^3 with V to obtain $p = nM\overline{v^2}/3V$.

☐ The **root-mean-square speed** v_{rms} of the molecules in a gas is defined by $v_{rms} = \sqrt{\overline{v^2}}$. Thus, $p = nMv_{rms}^2/3V$. Since $pV = nRT$,

$$v_{rms} = \sqrt{\frac{3RT}{M}}.$$

v_{rms} depends only on the temperature and the molar mass. It remains the same if the temperature does not change.

20–5 Translational Kinetic Energy

☐ For an ideal gas, the average translational kinetic energy per molecule is proportional to the temperature. In terms of the universal gas constant R and Avogadro's number N_A, the exact relationship is $\overline{K} = \frac{1}{2}mv_{rms}^2 = 3RT/2N_A$. The quantity $k = R/N_A$ is called the **Boltzmann constant** and has the value $k = 1.38 \times 10^{-23}$ J/K. In terms of k and

T,

$$\overline{K} = \frac{3}{2}kT .$$

☐ The molecules of two ideal gases in thermal equilibrium with each other have exactly the same average translational kinetic energy. The root-mean-square speeds, however, are different if the molecular masses are different. The average translational kinetic energy does not change if the temperature does not change.

20–6 Mean Free Path

☐ The **mean free path** λ of a molecule in a gas is the average distance it travels between collisions with other molecules. It can be calculated by considering the number of other molecules that are encountered by any given molecule as it moves through the gas. The result is

$$\lambda = \frac{1}{\sqrt{2}\pi d^2 N/V} ,$$

where d is the diameter of a molecule and N is the number of molecules in a volume V of the gas.

☐ The number of collisions suffered by a gas molecule per unit time is given by its average speed divided by its mean free path.

20–7 The Distribution of Molecular Speeds

☐ Not all molecules in a gas have the same speed. The distribution of speeds is described by the **Maxwell speed distribution** $P(v)$, defined so that $P(v)\,dv$ gives the fraction of molecules with speeds in the range from v to $v + dv$. If a gas at temperature T has molar mass M, then the Maxwell speed distribution is given by

$$P(v) = 4\pi \left(\frac{M}{2\pi RT} \right)^{3/2} v^2\, e^{-Mv^2/2RT} .$$

The function is graphed in Fig. 20–7 of the text. Notice that at low temperatures the function has a sharp, high peak while at high temperatures the peak is low and broad.

☐ To find the number of molecules with speed between v_1 and v_2, evaluate the definite integral

$$N(v_1, v_2) = \int_{v_1}^{v_2} P(v)\,dv\,.$$

If the interval $v_2 - v_1$ is small, you can approximate N by $P(v_1)(v_2 - v_1)$.

☐ The Maxwell speed distribution can be used to calculate the most probable speed, the average speed, and the root-mean-square speed. The **most probable speed** v_p is the one for which $P(v)$ is maximum. In terms of the molar mass and the temperature, it is given by

$$v_p = \sqrt{\frac{2RT}{M}}\,.$$

The **average speed** is given by

$$\overline{v} = \int_0^\infty v P(v)\,dv = \sqrt{\frac{8RT}{\pi M}}\,,$$

where the explicit expression for the Maxwell distribution was substituted for $P(v)$ and the integral was evaluated. The **mean-square speed** is given by

$$\overline{v^2} = \int_0^\infty v^2 P(v)\,dv = \frac{3RT}{M}\,,$$

and the root-mean-square speed v_{rms} is the square root of this:

$$v_{\mathrm{rms}} = \sqrt{\frac{3RT}{M}}\,.$$

☐ Note that all three speeds (v_p, $\overline{v}$, and v_{rms}) depend on the temperature as well as on the molar mass. When the temperature increases, they all increase. For a given gas at a given temperature, the greatest of the three characteristic speeds is v_{rms} and the smallest is v_P.

20–8 The Molar Specific Heats of an Ideal Gas

☐ For a *monatomic* ideal gas, the internal energy is just the total kinetic energy of the molecules. Since the average kinetic energy of a molecule is $\overline{K} = \frac{3}{2}kT$, the internal energy of n moles of an ideal gas is given by $E_{\text{int}} = \frac{3}{2}nN_AkT = \frac{3}{2}nRT$. The internal energy of an ideal gas depends only on the temperature and the amount of gas. For other systems, the internal energy may also depend on the pressure or other macroscopic quantities.

☐ Suppose that the temperature of a monatomic ideal gas is changed by ΔT while the volume is held constant. Then, the internal energy changes by $\Delta E_{\text{int}} = \frac{3}{2}R\,\Delta T$. Since no work is done, the entire increase or decrease in internal energy is due to heat absorbed or rejected and $Q = \frac{3}{2}R\,\Delta T$. The heat capacity at constant volume is $Q/\Delta T = 3nR/2$, and the molar specific heat at constant volume is $C_V = 3R/2$ (= 12.5 J/mol · K). It does not depend on temperature. In this chapter and the next, C is used to represent a *molar* specific heat, not a heat capacity.

☐ You should understand that $\Delta E_{\text{int}} = nC_V\Delta T$ for any process, whether it involves heat or not and whether the volume changes or not. However, $Q = nC_V\Delta T$ only for a constant volume process, during which no work is done. If, for example, work W is done by n moles of a monatomic ideal gas in thermal isolation (no heat transfer), then the change in internal energy is $\Delta E_{\text{int}} = -W$ and the change in temperature is $\Delta T = -W/nC_V$.

☐ The molar specific heat at constant pressure C_p is related to the molar specific heat at constant volume by $C_p = C_V + R$. More heat is absorbed at constant pressure than at constant volume for the same rise in temperature. Since the internal energy depends only on the temperature, its change is the same. The additional energy absorbed goes into the work done by the gas during the process.

20-9 Degrees of Freedom and Molar Specific Heats

☐ The **equipartition theorem** states that the internal energy is equally divided among the degrees of freedom of a system, with the energy associated with each degree of freedom being $\frac{1}{2}kT$ per molecule or $\frac{1}{2}RT$ per mole. One degree of freedom is associated with each independent energy term. For example, three degrees of freedom are associated with the translational kinetic energy of a molecule since it can move in any of three independent directions. Other degrees of freedom are associated with the rotational motions of the molecules.

☐ For a monatomic ideal gas, there are three degrees of freedom for each molecule, the molar specific heat at constant volume is $3R/2$, and the molar specific heat at constant pressure is $5R/2$. For a diatomic ideal gas, there are five degrees of freedom for each molecule (three translational and two rotational), the molar specific heat at constant volume is $5R/2$, and the molar specific heat at constant pressure is $7R/2$. For a polyatomic ideal gas, there are six degrees of freedom for each molecule (three translational and three rotational), the molar specific heat at constant volume is $3R$ and the molar specific heat at constant pressure is $4R$.

20-10 A Hint of Quantum Theory

☐ Classical theory predicts that the molar heat capacities of ideal gases should be independent of the temperature. In fact, they are not. If quantization of energy is taken into account, a decrease in the heat capacities with decreasing temperature is correctly predicted. According to quantum theory, the chance that the energy of a molecule is high diminishes rapidly as the temperature decreases, and this leads to a decrease in the heat capacities at lower temperatures.

☐ Atoms in diatomic and polyatomic molecules may also vibrate and the vibration modes may contribute to the heat capacities. Except at the highest temperatures, however, these are also frozen out.

20–11 The Adiabatic Expansion of an Ideal Gas

☐ As an ideal gas undergoes an adiabatic process the quantity pV^γ remains constant. Here γ is the ratio C_p/C_V. If p_i and V_i are the pressure and volume of the initial state and p_f and V_f are the pressure and volume of the final state then, $p_i V_i^\gamma = p_f V_f^\gamma$.

☐ If two of the three quantities p, V, and T are given for one state and one of them is given for another state, connected to the first by an adiabatic process, then $pV^\gamma =$ constant and $pV = nRT$ can be used to find values for the thermodynamic variables that have not been given.

NOTES:

Chapter 21
ENTROPY AND THE SECOND LAW
OF THERMODYNAMICS

Here you study the second law of thermodynamics and a closely associated property of any macroscopic system, its entropy. Understand the two definitions: in terms of heat and temperature and in terms of microstates of the system. Pay careful attention to calculations of entropy changes during various processes. Learn to distinguish between reversible and irreversible processes and understand how entropy behaves for each case. The second law has many far-reaching consequences; we can imagine many phenomena that do not violate any other laws of physics, including the conservation laws, but, nevertheless, do not occur. The second law gives the reason.

Important Concepts

☐ reversible process
☐ irreversible process
☐ quasi-static process
☐ entropy
☐ state variable
☐ second law of thermodynamics

☐ heat engine
☐ thermal efficiency
☐ refrigerator
☐ coefficient of performance
☐ microstate

21–1   Some One-Way Processes

☐   There are many phenomena that occur in one direction but not in the other: heat never flows from the cold end of a rod to the hot end, one part of an isolated object never spontaneously becomes hot while the other parts become cold, all of the air molecules in a room never spontaneously congregate in one corner. None of these phenomena are precluded by Newton's second law or the conservation laws. They are precluded by the **second law of thermodynamics**.

21–2 Change in Entropy

☐ A **quasi-static process** is one that is carried out so slowly that the system is essentially in thermodynamic equilibrium throughout. A quasi-static process without friction and turbulence is **reversible**; that is, it can be run backward with the signs of the heats and work changing but not their magnitudes. A process that is not reversible is said to be an **irreversible process**.

☐ If a system receives heat dQ as it goes from an initial equilibrium state to a nearby equilibrium state, then the **entropy** S of the system changes by $dS = dQ/T$, where T is the temperature. This defines entropy. If the state changes reversibly from some initial equilibrium state i to some final equilibrium state f, the change in entropy can be computed by evaluating an integral:

$$\Delta S = S_f - S_i = \int_i^f \frac{dQ}{T}.$$

The process connecting the initial and final states must be reversible for this equation to be valid.

☐ Entropy is a **state variable**. This means that whenever a system is in the same thermodynamic state, it has the same entropy, just as it has the same temperature, pressure, volume, and internal energy. The change in any of these variables over a cycle is zero.

☐ To evaluate the integral for ΔS, you must pick some *reversible* process that connects the initial and final states but, because entropy is a state variable, it need not be the process actually carried out on the system. Here are some examples:

1. Change in phase. Suppose mass m of a substance with heat of fusion L_F melts at temperature T. For this process, T is constant and $\Delta S = Q/T$. Since $Q = mL_F$,

$$\Delta S = \frac{mL_F}{T}.$$

The change in entropy on freezing is $\Delta S = -mL_F/T$. The entropy of the substance increases on melting and decreases on freezing.

2. Constant volume process. Suppose the molar specific heat at constant volume C_V for a certain gas is independent of T, V, and p. If an n-mole sample of this gas undergoes an infinitesimal reversible change of state at constant volume, so its temperature changes by dT, then the heat absorbed is $dQ = nC_V\,dT$ and the change in entropy is $dS = dQ/T = (nC_V/T)\,dT$. For a finite change in temperature, from T_1 to T_2, say, the change in entropy is the integral of this expression:

$$\Delta S = \int_{T_i}^{T_f} \frac{nC_V}{T}\,dT = nC_V \ln \frac{T_f}{T_i}\,.$$

3. Constant pressure process. Suppose the molar specific heat at constant pressure C_p for a certain gas is independent of T, V, and p. If an n-mole sample of this gas undergoes an infinitesimal reversible change of state at constant pressure, so its temperature changes by dT, then the heat absorbed is $dQ = nC_p\,dT$ and the change in entropy is $dS = dQ/T = (nC_p/T)\,dT$. For a finite change in temperature, from T_i to T_f, the change in entropy is the integral of this expression:

$$\Delta S = \int_{T_i}^{T_f} \frac{nC_p}{T}\,dT = nC_p \ln \frac{T_f}{T_i}\,.$$

4. Isothermal process. If a gas undergoes an infinitesimal reversible change of state at constant temperature, then $dE_{\text{int}} = nC_V\,dT = 0$ and $dQ = dW = p\,dV$. Thus, $dS = (p/T)\,dV$ and $\Delta S = \int(p/T)\,dV$. To evaluate this integral, p must be known as a function of volume and temperature. The equation of state gives this information. For

n moles of an ideal gas $p/T = nR/V$ and

$$\Delta S = \int_{V_i}^{V_f} \frac{nR}{V} \, dV = nR \ln \frac{V_f}{V_i} ,$$

where V_i is the initial volume and V_f is the final volume.

☐ Consider the adiabatic free expansion of an ideal gas from volume V_i to volume V_f, an irreversible process. No heat is received, no work is done, the change in the internal energy is zero, and the change in the temperature is zero. The entropy change can be calculated using a reversible isothermal expansion from V_i to V_f. It is $\Delta S = nR \ln(V_f/V_i)$. No heat is received but the entropy increases.

21–3 The Second Law of Thermodynamics

☐ The second law of thermodynamics is expressed in terms of entropy changes: In any thermodynamic process that proceeds from one equilibrium state to another, the total entropy of the system and its environment either remains unchanged or increases. If the process is reversible, the entropy is constant; if the process is irreversible, the entropy increases. In equation form, $dS \geq 0$.

☐ You must consider the *total* entropy of the system *and* its environment. Many processes lower the entropy of a system but, according to the second law, they must then increase the entropy of the environment.

21–4 Entropy in the Real World: Engines

☐ A heat engine takes heat from a high-temperature thermal reservoir, converts some of it to work, and rejects the remainder to a low-temperature thermal reservoir.

☐ A heat engine makes use of a working substance whose thermodynamic state varies during the process but that periodically returns to the same state. The same sequence of steps is performed on it during every cycle. If a heat engine receives heat of magnitude $|Q_H|$ from a high-temperature thermal reservoir and rejects heat of magnitude $|Q_C|$ to a low-temperature reservoir, then according to the first law of thermodynamics, it does work $W = |Q_H| - |Q_C|$ over a cycle. Note that because the engine works in cycles the change in the internal energy is zero.

☐ The **thermal efficiency** of a heat engine is given by

$$\varepsilon = \frac{|W|}{|Q_H|} = \frac{|Q_H| - |Q_C|}{|Q_H|} .$$

☐ A *perfect* heat engine takes in heat and does an identical amount of work. No heat is rejected. Since $|W| = |Q_H|$ for a perfect heat engine, its efficiency is 1. Such an engine does not violate the first law of thermodynamics but it does violate the second law. Over a cycle, the entropy of the high-temperature reservoir decreases by $|Q_H|/T_H$, the entropy of the low-temperature reservoir increases by $|Q_C|/T_C$, and the entropy of the working substance does not change. Thus, the change in the total entropy of the engine and its environment is

$$\Delta S = \frac{|Q_C|}{T_C} - \frac{|Q_H|}{T_H} .$$

According to the second law, this cannot be negative. That means $|Q_C|$ must be at least as great as $|Q_H|T_C/T_H$. In particular, it cannot be zero, so $|W|$ cannot be equal to $|Q_H|$ and the efficiency of a heat engine cannot be 1. The first law tells us that in a cycle heat input is turned to work and heat output. The second law tells us it cannot be turned completely to work; some must become heat output.

☐ A reversible engine is said to be *ideal*. The total entropy of such an engine and the reservoirs does not change. This means $|Q_C|/T_C = |Q_H|/T_H$ and the efficiency is

$$\varepsilon = \frac{T_H - T_C}{T_H} .$$

All reversible engines operating between the same temperatures have the same efficiency, regardless of the working substance and the thermodynamic processes involved, and this efficiency is the greatest possible for the temperatures involved.

☐ A Stirling engine provides a concrete example. The steps in a cycle are: an isothermal expansion at high temperature, a constant volume change in temperature to the lower temperature, an isothermal compression at low temperature, and a constant volume change in temperature back to the initial state. If the gas is ideal, the heat taken in during the increase in temperature at constant volume is the same in magnitude as the heat released during the decrease in temperature. This means we need to consider only the heat Q_H taken in during the high-temperature expansion and the heat Q_C released during the low-temperature compression. Suppose the working substance is n moles of a monatomic ideal gas and assume the engine is operated reversibly. Suppose the volume is expanded from V_i to V_f at temperature T_H and contracted from V_f to V_i at temperature T_C. Then, during a single cycle, the heat received from the high-temperature reservoir is given by $Q_H = W = \int p \, dV = nRT_H \ln(V_f/V_i)$, the heat rejected to the low-temperature reservoir is $Q_C = -nRT_C \ln(V_f/V_i)$, and the total work done is $W = nR(T_H - T_C) \ln(V_f/V_i)$. In terms of the temperatures of the reservoirs, the efficiency of the engine is $\varepsilon = W/Q_H = (T_H - T_C)/T_H$, in agreement with the general result obtained above.

21–5   Entropy in the Real World: Refrigerators

☐ A refrigerator takes in heat from a low-temperature reservoir and ejects heat to a high-temperature reservoir. Work must be done on the working substance of the refrigerator to accomplish this.

☐ Suppose that during a cycle a refrigerator takes in heat of magnitude $|Q_C|$ at a low-temperature reservoir and rejects heat of magnitude $|Q_H|$ at a high-temperature reservoir. Work of magnitude $|W| = |Q_H| - |Q_C|$ is done on the working substance of the refrigerator. The **coefficient of performance** is given by

$$K = \frac{|Q_C|}{|W|} = \frac{|Q_C|}{|Q_H| - |Q_C|}.$$

☐ For a *perfect* refrigerator $|W| = 0$ and K is infinite. A perfect refrigerator does not violate the first law of thermodynamics but it does violate the second law and cannot exist. Over a cycle, the change in the total entropy of the refrigerator and reservoirs is

$$\Delta S = \frac{|Q_H|}{T_H} - \frac{|Q_C|}{T_C}.$$

According to the second law, this cannot be negative. So $|Q_H|$ cannot be less than $|Q_C| T_H / T_C$ and the work done on the refrigerator cannot be less than $|Q_C|(T_H - T_C)/T_C$. Work must be done to cyclically absorb heat from a low-temperature reservoir and reject it to a high-temperature reservoir. The statement does NOT preclude heat flowing from a high-temperature reservoir to a low-temperature reservoir without work being done. This is a natural occurrence.

☐ An ideal refrigerator is one that is reversible. Over a cycle, the change in the total entropy of the refrigerator and reservoirs is zero. This means

$$\frac{|Q_H|}{T_H} = \frac{|Q_C|}{T_C}$$

and

$$K = \frac{T_C}{T_H - T_C}.$$

The coefficients of performance of all reversible refrigerators operating between the same two temperatures are the same and irreversible refrigerators have smaller coefficients.

21–6 A Statistical View of Entropy

☐ There are many different arrangements of atoms that lead to the same thermodynamic properties. All arrangements with the same properties are said to have the same configuration. Each arrangement is called a <u>microstate</u>. A different number of microstates may be associated with different configurations.

☐ The fundamental hypothesis of statistical mechanics is that every microstate, consistent with energy conservation, is equally probable. Thus, a system with a large number of particles will, with overwhelming probability, be in the configuration with the greatest number of microstates.

☐ The entropy of a system in a given configuration is related to the number of microstates associated with that configuration. If W is the number of microstates, then the relationship is $S = k \ln W$, where k is the Boltzmann constant.

NOTES:

Chapter 22
ELECTRIC CHARGE

You start your study of electromagnetism with Coulomb's law, which describes the force that one stationary charged particle exerts on another. Pay attention to both the magnitude and direction of the force and learn how to calculate the total force when more than one charge acts.

Important Concepts

☐ electric charge   ☐ Coulomb's law

☐ conductor   ☐ quantization of charge

☐ insulator   ☐ conservation of charge

22–2 Electric Charge

☐   **Electric charge** is a property possessed by some particles and, because they have charge, these particles exert electric forces on each other. The SI unit for charge is the coulomb (abbreviated C).

☐   There are two types of charge, called positive and negative. Charges of the same type repel each other; charges of different type attract each other. In many of the equations of electromagnetism, a positive number is substituted for the value of a positive charge and a negative number is substituted for the value of a negative charge.

☐   The charges on an electron and on a proton have exactly the same magnitude (1.60×10^{-19} C) but opposite signs. An electron is negative; a proton is positive.

☐   All macroscopic materials contain enormous numbers of negatively charged electrons and positively charged protons but if the number of electrons in an object equals the number of protons, then the net charge is zero and the object is said to be **neutral** (or uncharged). If there are more electrons than protons, the object has a net negative charge; if there are more protons than electrons, it has a net positive charge. In either case, it said to be **charged**. The net charge is computed by algebraically adding the charges of all particles in the object, taking their signs into account. Normally, macroscopic bodies are neutral.

☐ Objects may be given net charges by rubbing them together. When a glass rod is rubbed with silk, the rod becomes positively charged. When a plastic rod is rubbed with fur, the rod becomes negatively charged.

22–3 Conductors and Insulators

☐ The *electrons* in materials move and are transferred to or from other objects; the atomic nuclei (containing protons) are nearly immobile. Materials are often classified according to the freedom with which their electrons can move. Electrons in an **insulator** are not free to move; any charge placed on an insulator remains where it is placed. Scraping or rubbing is required to transfer charge to or from an insulator. **Conductors** contain many electrons that are free to move and easily flow between touching conductors.

☐ All metals, like copper, silver, and aluminum, are conductors. Rubber and mica are good insulators.

☐ A neutral conductor is attracted to a charged rod because the electrons within the inductor are redistributed, making some regions positively charged and others negatively charged. The forces on the negative and positive regions are in opposite directions but the force is greater on the region nearer to the charged object.

☐ Charged objects also attract neutral insulators but the force is much smaller than the force of attraction for a conductor. Under the influence of the external charge, electrons in an insulator move slightly from their normal orbits so the center of the negative charge is slightly displaced from the center of the positive charge. This results in a net force.

22–4 Coulomb's Law

☐ Suppose two point particles, one with charge of magnitude q_1 and the other with charge of magnitude q_2, are a distance r apart. The magnitude of the force exerted by either of the charges on the other is given by

$$F = \frac{1}{4\pi\epsilon_0} \frac{q_1 q_2}{r^2}.$$

The constant of nature $\epsilon_0 = 8.85 \times 10^{-12}$ $C^2 N./m^2$ is called the **permittivity constant**. The factor $k = 1/4\pi\epsilon_0$ has the value 8.99×10^9 N · m^2/C. The magnitude of the force is proportional to the inverse square of the distance between the charges and is also proportional to the product of the magnitudes of the charges.

☐ The direction of the electric force one charge exerts on another is along the line that joins the charges. If the charges have the same sign, the force on either is away from the other charge; if they have different signs, the force is toward the other charge. Electric forces between two charges obey Newton's third law: they are equal in magnitude and opposite in direction.

☐ When more than two charges are present, the force on any one of them is the *vector* sum of the forces due to the others. Each force is computed using Coulomb's law. This is the *principle of superposition* for electric forces.

☐ Just as for gravitational forces there are two shell theorems for electrical forces. They are:

The electrical force between a uniformly charged spherical shell and a point charge *outside* the shell is as if the charge of the shell were concentrated at its center.

The electrical force between a uniformly charged spherical shell and point charge anywhere *inside* the shell is zero.

22–5 Charge is Quantized

☐ The charge on all particles detected so far is a positive or negative multiple of a fundamental unit of charge e (1.60×10^{-19} C).

22–6 Charge is Conserved

☐ The sum of the charges (including signs) on all particles in a closed system remains the same. This is true when the particles do not change identities, as in atomic physics and chemistry. It is also true in the subatomic world, where particles can disappear and others can appear. For example, an electron and positron can annihilate each other, leaving two uncharged quanta of light (photons). The charge on the electron is $-e$ and the charge on the positron is $+e$, so the total charge is zero, both before and after the annihilation event.

NOTES:

Chapter 22:   Electric Charge

Chapter 23
ELECTRIC FIELDS

The idea of an electric field is introduced and used to describe electrical interactions between charges. It is fundamental to our understanding of electromagnetic phenomena and is used extensively throughout the rest of this course.

Important Concepts

☐ field

☐ electric field

☐ electric field lines

☐ electric field of a point charge

☐ linear charge density

☐ surface charge density

☐ electric dipole

☐ electric field of a dipole

☐ torque on a dipole

☐ potential energy of a dipole

23–1 Charges and Forces: A Closer Look

☐ A **field** gives some property of a region as a function of position. Temperature, for example, is a scalar field and the particle velocity as a function of position in a fluid is a vector field.

☐ **Electric fields** are associated with electric charges. The electric field associated with a charge pervades all space and exerts a force on any other charge in it.

23–2 The Electric Field

☐ The electric field at any point in space is defined as the electric force per unit test charge on a positive test charge placed at that point, in the limit as the test charge becomes vanishingly small. If the test charge is q_0 and $\mathbf{F}$ is the electric force on it, then the electric field at the position of the test charge is $\mathbf{E} = \mathbf{F}/q_0$. $\mathbf{E}$ is the total field created by all charges other than the test charge. An electric field is associated with the charges that create it and exists regardless of the presence of a test charge. The SI units of an electric field are N/C.

☐ The limit as q_0 becomes vanishingly small must be included in the definition to minimize the influence of the test charge on the charges creating the field being measured.

☐ If a collection of stationary charges creates an electric field **E** at the position of another charge q, then the force exerted by the field on q is given by $\mathbf{F} = q\mathbf{E}$.

23–3 Electric Field Lines

☐ **Electric field lines** (also called lines of force) graphically depict both the direction and magnitude of an electric field. At any point the field is tangent to the line through that point. The lines are more concentrated in regions of high electric field than in regions of low electric field. In fact, the number of lines per unit area passing through a small area perpendicular to the lines is proportional to the magnitude of the field. Arrows are placed on field lines to indicate the direction of the field but electric field lines themselves are *not* vectors. Electric field lines emanate from positive charge and terminate on negative charge.

☐ To see the electric field lines for some charge distributions, look carefully at Figs. 23–2, 3, 4, and 5 of the text. For each figure notice the directions of the lines, where they are close together, and where they are far apart.

23–4 The Electric Field Due to a Point Charge

☐ The magnitude of the electric field at a point a distance r from a single point charge q is given by $E = q/4\pi\epsilon_0 r^2$. The field is along the line that joins the charge q and the point. If q is positive, it points away from q; if q is negative, it points toward q.

□ If more than one point charge is responsible for the electric field, the total field is the *vector* sum of the fields due to the individual charges.

23–5 The Electric Field Due to an Electric Dipole

□ **An electric dipole** consists of a positive charge $+q$ and a negative charge $-q$ (of equal magnitude), separated by a distance d. It is characterized by a dipole moment **p**, a vector with magnitude given by $p = qd$. The direction of **p** is along the line joining the charges, from the negative charge toward the positive charge. The field lines of a dipole are shown in Fig. 23–5.

□ In Fig. 23–8, the z axis is in the direction of the dipole moment and the origin is midway between the charges. The magnitude of the field produced at P by the positive charge is $E_+ = q/4\pi\epsilon_0(z - d/2)^2$ and the magnitude of the field produced there by the negative charge is $E_- = q/4\pi\epsilon_0(z + d/2)^2$. The first field is in the positive z direction and the second is in the negative z direction. The total field is

$$E = \frac{q}{4\pi\epsilon_0} \left[\frac{1}{(z - d/2)^2} - \frac{1}{(z + d/2)^2} \right] .$$

It is in the positive z direction.

□ Usually we are interested in situations for which $z \gg d$. The binomial theorem can be used to expand $(z - d/2)^2$ and $(z + d/2)^2$. When the only terms retained are those that are proportional to d, the expression for the field becomes $E = p/2\pi\epsilon_0 z^3$. The field is inversely proportional to z^3, not z^2.

23–6 The Electric Field Due to a Line of Charge

□ A continuous distribution of charge is characterized by a **charge density**. If the charge is on a line (either straight or curved), the appropriate charge density is the **linear charge density**, denoted by λ and defined so that the charge dq in an infinitesimal segment ds of the line is $dq = \lambda\,ds$. If the charge is distributed on a surface, the appropriate charge density is the **surface charge density**, denoted by σ and defined so that the charge dq in an infinitesimal element of area dA is $dq = \sigma\,dA$. For a *uniform* charge distribution, the appropriate charge density is the same everywhere along the line or on the area.

☐ To calculate the electric field produced by a uniform ring of charge at a point along its axis, take the z axis to coincide with the axis of the ring and place the origin at the center of the ring. See Fig. 23–9 of the text. Two diametrically opposite elements of the ring produce fields with horizontal components that are the negatives of each other and that sum to zero. The total field is thus in the z direction. An infinitesimal segment ds of the ring has charge $dq = \lambda\,ds$, where λ is the linear charge density. The magnitude of the field produced by the segment at P is $E = (1/4\pi\epsilon_0)(\lambda/r^2)\,ds$, where r is the distance from the segment to P. It is given by $r = (z^2 + R^2)^{1/2}$, where R is the radius of the ring. To obtain the z component, multiply the magnitude of the field by the cosine of the angle θ between the field and the z axis. This is given by $\cos\theta = z/r = z/(z^2 + R^2)^{1/2}$. The total field is found by integrating around the ring. All quantities in the integrand have the same value for all segments of the ring and ds integrates to $2\pi R$. Thus,

$$E = \frac{z\lambda(2\pi R)}{4\pi\epsilon_0(z^2 + R^2)^{3/2}} = \frac{qz}{4\pi\epsilon_0(z^2 + R^2)^{3/2}}.$$

23–7 The Electric Field Due to a Charged Disk

☐ To calculate the field produced by a uniform disk of charge, the disk is divided into rings, each with infinitesimal width dr. The area of a ring of radius r is $dA = 2\pi r\,dr$ and if σ

is the area charge density, then $dq = 2\pi r\sigma\,dr$ is the charge in such a ring. The field it produces at a point on the axis a distance z from the ring center is along the axis and has the magnitude

$$dE = \frac{z\sigma(2\pi r)}{4\pi\epsilon_0(z^2 + R^2)^{3/2}}\,dr\,.$$

To find the total field, this expression is integrated from $r = 0$ to $r = R$. The result is:

$$E = \frac{\sigma}{2\epsilon_0}\left(1 - \frac{z}{\sqrt{z^2 + R^2}}\right)\,.$$

☐ An expression for the field produced by a uniform plane of charge can be found by allowing R to be large without bound. If $R \gg z$, then $z/\sqrt{z^2 + R^2}$ is much smaller than 1 and $E = \sigma/2\epsilon_0$.

☐ At points near a ring, disk, or plane, superposition produces total fields that are quite different from each other and quite different from the field of a point charge. At points far from a charge distribution, however, the electric field tends to become like that of a point particle with charge equal to the net charge in the distribution.

23–8   A Point Charge in an Electric Field

☐ If an electric force is the only force acting on a charge, then Newton's second law takes the form $q\mathbf{E} = m\mathbf{a}$, where $\mathbf{a}$ is the acceleration of the charge and m is its mass. If you know the initial position and velocity of the charge, then you can, in principle, predict its subsequent motion. If the field is uniform, use the kinematic equations for constant acceleration.

23–9 A Dipole in an Electric Field

☐ The net force on a dipole in a uniform electric field is zero because the individual forces on the two charges have the same magnitude and are opposite in direction.

☐ A uniform electric field exerts a torque on a dipole. If **p** is the dipole moment and **E** is the electric field, then the torque is $\boldsymbol{\tau} = \mathbf{p} \times \mathbf{E}$. This torque tends to rotate the dipole so that **p** becomes more closely aligned with the field.

☐ When an electric dipole rotates in an electric field **E**, the field does work on it. If the angle between the dipole moment and field changes from θ_0 to θ, the work done by the field is

$$W = \int_{\theta_0}^{\theta} \tau \, d\theta = -\int_{\theta_0}^{\theta} pE \sin\theta \, d\theta = pE(\cos\theta - \cos\theta_0).$$

The sign of the work is negative if the angle increases and positive if the angle decreases.

☐ When a dipole rotates in an electric field, the potential energy changes by $\Delta U = -W = -pE(\cos\theta - \cos\theta_0)$. The potential energy for any orientation can be written $U = -pE\cos\theta = -\mathbf{p} \cdot \mathbf{E}$. It is a maximum when $\theta = 180°$ and is a minimum when $\theta = 0$.

NOTES:

Chapter 24
GAUSS' LAW

Gauss' law is one of the four fundamental laws of electromagnetism. It relates an electric field to the charges that create it and is, therefore, closely akin to Coulomb's law. Unlike Coulomb's law, however, it is valid when the charges are moving, even at relativistic speeds. The central concept for an understanding of Gauss' law is that of electric flux, a quantity that is proportional to the number of field lines penetrating a given surface. Pay careful attention to the definition of flux and learn how to compute it for various fields and surfaces. Then, learn how to use Gauss' law to compute the charge in any region if the electric field is known on the boundary and also how to use the law to compute the electric field in certain highly symmetric situations.

Important Concepts

☐ flux ☐ Gauss' law
☐ electric flux ☐ Gaussian surface

24–1 A New Look at Coulomb's Law

☐ To use Gauss' law, you consider a *closed* surface, called a **Gaussian surface**. It can be a real surface, such as the boundary of a material object, or an imaginary surface, constructed only for the purpose of applying the law. When Gauss' law is used to solve for the electric field, the Gaussian surface should reflect the symmetry of the problem: a sphere for a point charge or a spherical conductor, a cylinder for a wire or a cylindrical conductor.

☐ Gauss' law gives the relationship between the *normal* component of the electric field at a Gaussian surface and the electric charge enclosed by the surface. This component is directly related to the **flux** of the electric field.

24–2 Flux

☐ For a moving fluid, the flux through an area is the volume of fluid that crosses the area per unit time per unit area. If the fluid velocity vector **v** makes the angle θ with the normal to the surface, the flux is $\Phi = (v \cos \theta)A$, where A is the area.

☐ $\Phi = 0$ if the velocity is parallel to the surface ($\theta = 90°$). No fluid moves through the surface then. Φ has its maximum value if the velocity is perpendicular to the surface ($\theta = 0$).

24–3 Flux of an Electric Field

☐ Although the electric field does not represent a velocity, a flux can be associated with it. If the field is uniform, the electric flux through a plane area A is given by $\Phi = (E \cos \theta)A$, where θ is the angle between the field and the normal to the area. If the field is not uniform, the flux is defined by the integral $\Phi = \int (E \cos \theta)\, dA$. The surface is divided into a large number of small area elements, the flux through each element is calculated, and the results are summed.

☐ If the *vector* area element d**A** is defined to be in the direction of the normal to the surface, the electric flux can be written $\Phi = \int \mathbf{E} \cdot d\mathbf{A}$.

☐ The vector area element d**A** may be in either of the two directions that are perpendicular to the surface (up or down for a horizontal surface). Which one is used determines the sign but not the magnitude of the flux. If the surface is closed, like a Gaussian surface, d**A** is always chosen to point *outward* and the definition of electric flux is written $\Phi = \oint \mathbf{E} \cdot d\mathbf{A}$, with a small circle on the integral sign to indicate that the surface is closed.

☐ An area element contributes to the electric flux only if the field pierces the element. In fact, the flux through a surface is proportional to the number of field lines that penetrate the element. Field lines that cross a closed surface from inside to outside make a positive contribution to Φ while lines that cross the surface from outside to inside make a negative contribution. Since some parts of a closed surface may make positive contributions to the flux while other parts make negative contributions, the total flux through a surface may vanish even though an electric field exists at every point on the surface.

☐   Remember that the number of electric field lines per unit area through an infinitesimal area dA, perpendicular to the field, is proportional to the magnitude of the field. If the vector $d\mathbf{A}$ is in the same direction as the field, the number of lines that penetrate it is proportional to $E\,dA$. If the area is rotated so $d\mathbf{A}$ makes the angle θ with the field, the number of lines that penetrate the area decreases to $E\cos\theta\,dA$.

24–4   Gauss' Law

☐   Gauss' law states that the total electric flux through any closed surface is proportional to the total charge enclosed by the surface. Mathematically,

$$\epsilon_0 \oint \mathbf{E}\cdot d\mathbf{A} = q\,,$$

where q is the total charge enclosed, the sum of the charge (including sign) on all particles within the surface.

☐   The law should not surprise you since the magnitude of the electric field and the number of field lines associated with any charge are both proportional to the charge. All lines from a single positive charge within a closed surface penetrate the surface from inside to outside, so the total flux is positive and proportional to the charge. All lines associated with a single negative charge within a closed surface penetrate from outside to inside, so the total flux is negative and again proportional to the charge. Some lines associated with a charge outside a surface penetrate the surface, but those that do, penetrate twice, once from outside to inside and once from inside to outside, so this charge does not contribute to the total flux through the surface.

24-5 Gauss' Law and Coulomb's Law

☐ Gauss' law can be used to find an expression for the electric field of a point charge. Imagine a sphere of radius r with a positive point charge q at its center. Since the electric field is radially outward from the charge, the normal component of the field at any point on the surface of the sphere is the same as the magnitude E of the field. Furthermore, the magnitude E is uniform on the surface, so the total flux through the sphere is $\Phi = 4\pi r^2 E$. Equate this to q/ϵ_0 and obtain $E = q/4\pi\epsilon_0 r^2$, in agreement with Coulomb's law.

☐ If the electric field is given at all points on a surface, Gauss' law can be used to calculate the net charge enclosed by the surface. If the charge is known, the law can be used to find the total electric flux through a surface. If the charge distribution is highly symmetric, so a symmetry argument can be used to show that $E\cos\theta$ has the same value at all points on a surface, then Gauss' law can be used to solve for the electric field at points on the surface.

24-6 A Charged Isolated Conductor

☐ The electric field vanishes at all points in the interior of a conductor in electrostatic equilibrium, with all charge stationary. If this were not true, the electric field would exert a force on the electrons in the conductor and they would move until the force on each of them is zero. $E = 0$ inside a conductor even when excess charge is placed on it or when an external field is applied to it. In the interior, the external field is canceled by the field produced by the redistributed charge on the surface.

☐ In an electrostatic situation, any excess charge on a conductor must reside on its surface; there can be no net charge in its interior. Imagine a Gaussian surface that is completely within the conductor. The electric field is zero at every point on the surface, so the total flux through the surface is zero and, according to Gauss' law, the net charge enclosed by the surface is zero. Any net charge on the conductor must lie outside the Gaussian surface. Since this result is true for *every* Gaussian surface that can be drawn completely within the conductor, no matter how close to its surface, we conclude that any excess charge must be on the surface of the conductor. If the object is an insulator, the field may not be zero inside and excess charge may be distributed throughout its volume.

☐ If a conductor has a cavity (completely surrounded by the conductor), then the charge on the inside surface of the conductor (its boundary with the cavity) must be the negative of the charge in the cavity. According to Gauss' law, the sum of the charge in the cavity and on the inside surface must be zero. If there is charge Q in the cavity, then there is charge $-Q$ on the inside surface of the conductor. The charge on the inside and outside surfaces must sum to give the total charge on the conductor. If, for example, the total charge on the conductor is zero, then the charge on its outside surface is $+Q$.

☐ The magnitude of the electric field at any point just outside the surface of a conductor is directly proportional to the surface charge density at the corresponding point on the surface. The exact relationship is $E = \sigma/\epsilon_0$, where σ is the area charge density. This result follows directly from Gauss' law, applied to a small closed surface that is partially inside and partially outside the conductor.

24–7 Applying Gauss' Law: Cylindrical Symmetry

☐ Successful use of Gauss' law to solve for the electric field depends greatly on choosing the right Gaussian surface. First, it must pass through the point where you want the value of the field. Second, either the normal component of the electric field must have constant magnitude over the entire surface or else it must have constant magnitude over part of the surface and the flux through the other parts must be zero. A symmetry argument should be made to justify the use of the Gaussian surface you have chosen.

☐ If a long, straight, cylindrical rod has positive charge distributed uniformly along it, symmetry tells us that the electric field is radially outward from the rod and has the same magnitude at points that are the same distance from the rod.

☐ To calculate the electric field a distance r from a uniformly charged rod, use a cylindrical Gaussian surface with radius r and length ℓ, concentric with the rod. The field has the same magnitude at all points on the rounded portion of the cylinder and at all these points it is normal to the surface. Over the rounded portion of the cylinder, $\int \mathbf{E} \cdot d\mathbf{A} = 2\pi E r \ell$. The electric field is parallel to the ends of the cylinder (perpendicular to $d\mathbf{A}$), so $\int \mathbf{E} \cdot d\mathbf{A} = 0$ for these portions of the Gaussian surface. Thus, $\oint \mathbf{E} \cdot d\mathbf{A} = 2\pi E r \ell$. The charge enclosed is $q = \lambda \ell$, where λ is the linear charge density of the rod. Gauss' law becomes $2\pi\epsilon_0 E r \ell = \lambda \ell$, so

$$E = \frac{\lambda}{2\pi\epsilon_0 r} \, .$$

24–8 Applying Gauss' Law: Planar Symmetry

☐ If an infinite sheet has a uniform, positive surface charge density σ, symmetry tells us that the electric field points perpendicularly away from the sheet and that the magnitude of the field is uniform over any plane that is parallel to the sheet. Take the Gaussian surface to be a cylinder with its ends parallel to the sheet and its axis perpendicular to the sheet. The sheet cuts through the midpoint of the cylinder's length, so the field has the same magnitude on both ends of the cylinder. The flux through the rounded portion of the cylinder is zero because the field is parallel to this portion of the surface (perpendicular to $d\mathbf{A}$); the flux through either of the cylinder ends is $\int \mathbf{E} \cdot d\mathbf{A} = EA$, where A is the area of an end. The charge enclosed by the Gaussian surface is $q = \sigma A$, so Gauss' law becomes $2\epsilon_0 EA = \sigma A$ and

$$E = \frac{\sigma}{2\epsilon_0}.$$

☐ You might think this result is inconsistent with the expression for the magnitude of the electric field just outside a conductor, $E = \sigma/\epsilon_0$. It is not. Consider an infinite plane conducting sheet with a uniform charge density σ on one surface. This charge produces an electric field with magnitude $\sigma/2\epsilon_0$, just like any other large uniform sheet of charge. The field exists on both sides of the sheet, in the interior of the conductor as well as in the exterior, and on each side it points away from the surface if σ is positive. Another electric field must be present, produced perhaps by charge on another portion of the conductor or by external charge. In the interior of the conductor, the second field exactly cancels the field due to the charge layer to produce a total field of zero and, in the exterior, it augments the field due to the charge layer to produce a total field with magnitude σ/ϵ_0.

24–9 Applying Gauss' Law: Spherical Symmetry

☐ If a spherical shell is uniformly charged, we use a Gaussian surface in the form of a sphere, concentric with the shell. The electric field is normal to the surface and its magnitude is uniform over the surface. The flux through the surface is $\Phi = 4\pi r^2 E$, where r is the radius of the Gaussian sphere. If the Gaussian surface is inside the shell, the charge enclosed is zero, so the electric field is $E = 0$ at points inside the shell. If the Gaussian surface is outside the shell, the charge enclosed is the total charge Q on the shell, so the electric field is $E = Q/4\pi\epsilon_0 r^2$ at points outside the shell.

☐ Consider a uniform sphere of charge with radius R. The field inside the sphere, a distance r (with $r < R$) from its center, is $E = Q'/4\pi\epsilon_0 r^2$, where Q' is the charge enclosed by a sphere of radius r. If the volume charge density is ρ, then $Q' = (4\pi/3)\rho r^3$ and

$$E = \frac{\rho r}{3\epsilon_0}.$$

If Q is the total charge in the sphere, then $\rho = 3Q/4\pi R^3$ and

$$E = \frac{Qr}{4\pi\epsilon_0 R^3}.$$

NOTES:

Chapter 25
ELECTRIC POTENTIAL

Because the electric force is conservative, a potential energy is associated with a collection of charges. If the charges are released, potential energy is converted to kinetic energy as each charge moves in response to the forces of the other charges. Electric potential is closely related to potential energy and plays a vital role in most succeeding discussions of electricity. Play careful attention to its definition and learn how to compute it for a collection of point charges.

Important Concepts

☐ electric potential   ☐ electric potential energy
☐ equipotential surface

25–1 Electric Potential Energy

☐ Since the electric force is conservative, a potential energy is associated with it, just as a potential energy is associated with the gravitational force. If a system of charges changes from some initial configuration to some final configuration, the change in the potential energy is given by $\Delta U = U_f - U_i = -W_{if}$, where W_{if} is the work done by electrical forces.

☐ A special case is of great importance: a test charge is moved in the electric field created by other charges, which remain stationary. The potential energy of the system consisting of the test charge and the other charges is the negative of the work done by the electrical forces acting on the test charge. If the test charge is moved from a reference point (usually far away from all other charges) and the potential energy is taken to be zero for the test charge at that point, then the negative of the work is the potential energy for the test charge at its final position.

25–2 Electric Potential

☐ Electric potential and electric potential energy are closely related but they are not the same. To find the electric potential of a collection of point charges, a reference point is chosen and the electric potential at that point is set equal to zero. Then, a positive *test* charge, not one of the charges in the collection, is moved from the reference point to any point P and the work done by the electric field on the test charge is calculated. The electric potential at P is the negative of this work divided by the test charge. This is also the change in the electric potential energy per unit test charge of the system consisting of the original collection of charges and the test charge. The charges of the collection must remain in fixed positions as the test charge is moved. If the total charge is finite, the reference point is usually selected to be infinitely far removed from the charge collection. A value for the electric potential is associated with each point in space and exists regardless of whether a test charge is present.

☐ Since electric potential is an energy divided by a charge, its SI unit is J/C. This unit is called a volt (abbreviation: V). The unit of an electric field may be taken to be V/m, which is the same as N/C.

25–3 Equipotential Surfaces

☐ An equipotential surface is a surface (imaginary or real) such that the electric potential has the same value at all points on it. It can be labelled by giving the value of the potential on it. Equipotential surfaces do not cross each other. One and only one goes through any point in space.

☐ The electric field line through any point is perpendicular to the equipotential surface through that point. If the field has a non-vanishing component tangent to a surface, then a potential difference must exist between points on the surface and the surface cannot be an equipotential surface.

☐ The equipotential surfaces of an isolated point charge are spheres, centered on the charge. The equipotential surfaces of a uniform field are planes, perpendicular to the field. Equipotential surfaces associated with other charge distributions are more complicated but if the distribution has a net charge, they are nearly spheres far from the distribution. Look at Fig. 25–3 of the text to see the surfaces associated with a dipole, for which the net charge is zero. All conductors are equipotential volumes and their boundaries are equipotential surfaces.

25–4 Calculating the Potential from the Field

☐ When a test charge q_0 moves through an infinitesimal displacement ds, the work done by the electric field is $dW = q_0\mathbf{E} \cdot \mathbf{ds}$ and the change in the potential energy is $dU = -dW = -q_0\mathbf{E} \cdot \mathbf{ds}$. The difference in the potential energy for two points i and f is given by

$$V_f - V_i = -\int_i^f \mathbf{E} \cdot \mathbf{ds}.$$

The potential at any point is

$$V = -\int_i^f \mathbf{E} \cdot \mathbf{ds},$$

where i is taken to be the reference point and the potential is taken to be zero there. These integrals are line integrals: the path is divided into infinitesimal segments, the integrand is evaluated for each segment, and the results are summed. Because the electric field is conservative, every path will give the same value for the potential difference of the given points.

☐ If the field is uniform, as it is outside a large sheet with uniform charge density, then $V_f - V_i = -\mathbf{E} \cdot \Delta\mathbf{r}$, where $\Delta\mathbf{r}$ is the displacement of point f from point i.

☐ Roughly speaking, the electric field points from a region where the potential is high toward a region where the potential is low. A positive charge is accelerated toward a low potential region; a negative charge is accelerated toward a high potential region.

25–5 Potential Due to a Point Charge

☐ As a test charge q_0 moves from far away to a point a distance r from a single isolated point charge q, the work done by the electric field on the charge is $W = -qq_0/4\pi\epsilon_0 r$ and the change in the potential energy of the two-charge system is $\Delta U = +qq_0/4\pi\epsilon_0 r$. If the potential is taken to be zero at infinity, then its value at the final position of the test charge is

$$V = \frac{q}{4\pi\epsilon_0 r}.$$

This expression is also valid if q is negative. The potential then has a negative value.

25–6 Potential Due to a Group of Point Charges

☐ Electric potential is a scalar, so the potential of a collection of point charges is the algebraic sum of the individual potentials of the charges in the collection.

☐ Suppose the collection consists of charges q_1, q_2, and q_3. The electric potential at some point P is

$$V = \frac{1}{4\pi\epsilon_0} \left(\frac{q_1}{r_1} + \frac{q_2}{r_2} + \frac{q_3}{r_3} \right),$$

where r_1 is the distance from q_1 to P, r_2 is the distance from q_2 to P, and r_3 is the distance from q_3 to P. When you substitute values for the charges, be sure you include the appropriate signs.

25–7   Potential Due to an Electric Dipole

☐ An electric dipole consists of a positive charge and a negative charge of the same magnitude q, separated by a distance d. The potential at any point can be found by summing the potentials of the two charges. Let $\mathbf{r}$ be the vector from the midpoint of the line joining the charges to the point and suppose this line makes the angle θ with the dipole moment. If the charges are close together compared to r, the potential is $V = p\cos\theta/4\pi\epsilon_0 r^2$.

25–8   Potential Due to a Continuous Charge Distribution

☐ Charge may be distributed continuously along a line, on a surface, or throughout a volume. If it is, divide the region into infinitesimal elements, each containing charge dq, and sum (integrate) the potential due to the regions. If an infinitesimal region is a distance r from point P, then the contribution of that region to the potential at P is $dV = dq/4\pi\epsilon_0 r$ and the potential due to the whole distribution is

$$V = \frac{1}{4\pi\epsilon_0}\int \frac{dq}{r}\, .$$

In practice, dq is replaced by $\lambda\, ds$ for a line distribution and by $\sigma\, dA$ for a surface distribution. Here λ is a linear charge density and σ is a surface charge density.

☐ Consider a uniform line of charge that extends from $x = 0$ to $x = L$, as shown in Fig. 25–13. If the linear charge density is λ, then the charge in an infinitesimal segment dx is $dq = \lambda\, dx$. If the coordinate of the segment is x and you are calculating the potential at a point that is a perpendicular distance d from the end of the line at $x = 0$, then the distance from the segment to the point is $r = (x^2 + d^2)^{1/2}$. The potential at the point due to the segment is

$$dV = \frac{1}{4\pi\epsilon_0}\frac{\lambda}{(x^2 + d^2)^{1/2}}\, dx$$

and the potential due to the entire line is

$$V = \frac{\lambda}{4\pi\epsilon_0} \int_0^L \frac{\mathrm{d}x}{(x^2 + d^2)^{1/2}}$$

$$= \frac{\lambda}{4\pi\epsilon_0} \ln\left(\frac{L + (L^2 + d^2)^{1/2}}{d}\right) .$$

☐ To calculate the potential produced by a uniform disk of charge at a point on the axis a distance z above the disk center, first consider a ring with radius R' and width $\mathrm{d}R'$. The area of the ring is $\mathrm{d}A = 2\pi R' \mathrm{d}R'$ and if σ is the surface charge density, the charge contained in the ring is $\mathrm{d}q = 2\pi\sigma R' \mathrm{d}R'$. All charge in the ring is the same distance from the point: $r = (z^2 + R'^2)^{1/2}$. Thus, the potential produced by the ring at z is

$$\mathrm{d}V = \frac{1}{4\pi\epsilon_0} \frac{\sigma(2\pi R')}{(z^2 + R'^2)^{1/2}} \mathrm{d}R'$$

and the total potential due to the disk is given by

$$V = \frac{\sigma}{2\epsilon_0} \int_0^R \frac{\mathrm{d}R'}{(z^2 + R'^2)^{1/2}} = \frac{\sigma}{2\epsilon_0}\left(\sqrt{z^2 + R^2} - z\right) .$$

25-9 Calculating the Field from the Potential

☐ If the electric potential is a known function of position in a region of space, then the components of **E** can be found:

$$E_x = -\frac{\partial V}{\partial x}, \qquad E_y = -\frac{\partial V}{\partial y}, \qquad \text{and} \qquad E_z = -\frac{\partial V}{\partial z} .$$

When you differentiate with respect to x to find E_x, you treat y and z as constants. Similar statements hold for E_y and E_z.

25–10 Electric Potential Energy of a System of Point Charges

□ Suppose charges q_1 and q_2 start a distance r_i apart and move so they end a distance r_f apart. Then, the work done by the electric field is

$$W_{if} = -\frac{q_1 q_2}{4\pi\epsilon_0}\left[\frac{1}{r_f} - \frac{1}{r_i}\right]$$

and the change in the potential energy of the two-charge system is

$$\Delta U = +\frac{q_1 q_2}{4\pi\epsilon_0}\left[\frac{1}{r_f} - \frac{1}{r_i}\right] .$$

The result is the same if either charge remains stationary while the other moves or if both move. All that matters is their initial and final separations.

□ If the total charge in the collection is finite, the potential energy is usually taken to be zero for infinite charge separation. Let the initial separation r_i become large without bound and replace r_f with r. Then,

$$U = \frac{q_1 q_2}{4\pi\epsilon_0 r}$$

gives the potential energy when their separation is r. This expression is valid no matter what the signs of the charges. If they have the same sign, the potential energy is positive; it decreases if their separation becomes greater and increases if their separation becomes less. If the two charges have opposite signs, the potential energy is negative; it increases (becomes less negative) if their separation becomes greater and decreases (becomes more negative) if their separation becomes less.

□ To calculate the potential energy of a collection of point charges, sum the potential energies of all *pairs* of charges. If the system consists of four charges, for example, add the potential energies of q_1 and q_2, q_1 and q_3, q_1 and q_4, q_2 and q_3, q_2 and q_4, and q_3 and q_4.

☐ The electric potential energy of a system is the work that must be done by an external agent to assemble the collection, bringing the charges from infinite separation to their final positions. The charges are at rest at the beginning and end of the process, so there is no change in kinetic energy.

☐ If no external agent acts on a system of charges, its mechanical energy is conserved: during any interval $\Delta K + \Delta U = 0$. If the charges are released from some initial configuration, their mutual attractions and repulsions might cause them to move. The potential energy of the system will decrease as it is converted to kinetic energy.

25–11 Electric Potential of a Charged Isolated Conductor

☐ All points in the interior and on the surface of an isolated conductor have the same electric potential, once equilibrium is established. This follows immediately from the definition of the potential difference between two points and the condition $\mathbf{E} = 0$ inside a conductor in equilibrium.

☐ Consider a conducting sphere of radius R, with charge Q uniformly distributed on its surface. If r is the distance from the sphere center to a point outside the sphere, the electric potential there is given by $V = Q/4\pi\epsilon_0 r$. If r is the distance from the sphere center to a point inside the sphere, the potential there is given by $V = Q/4\pi\epsilon_0 R$. The potential inside does not depend on r and has the same value as the potential at the surface. The zero of potential was taken to be at $r = \infty$.

NOTES:

Chapter 26
CAPACITANCE

Capacitors are electrical devices that are used to store charge and electrical energy and, as you will see in a later chapter, are important for the generation of electromagnetic oscillations. This chapter is an excellent review of the principles you have learned in previous chapters: you will make extensive use of Gauss' law and the concepts of electric potential and electric potential energy.

Important Concepts

□ capacitor
□ capacitance
□ equivalent capacitance

□ energy in a capacitor
□ electric energy density
□ dielectric

26–2   Capacitance

□ A **capacitor** is simply two conductors, isolated from each other. Each is called a **plate**. The symbol used to represent a capacitor in an electrical circuit is ⊣ ⊢.

□ In normal use, one plate holds positive charge and the other holds negative charge of the same magnitude. The charge creates an electric field, pointing roughly from the positive plate toward the negative plate, and because an electric field exists in the region between the plates, the positive plate is at a higher electric potential than the negative.

□ The potential difference of the plates and the charge on either one are proportional to each other. If the magnitude of the potential difference is V and the magnitude of the charge on either plate is q, then $q = CV$ defines the **capacitance** C of the capacitor. The capacitance is independent of both q and V. It does depend on the geometry of the capacitor. A capacitor can be charged by transferring charge from one plate to the other. The capacitance is a measure of how much is transferred for a given potential difference.

☐ A capacitor can be charged by connecting one plate to the positive terminal of a battery and the other plate to the negative terminal. The battery pulls electrons from one plate and deposits electrons on the other until the potential difference between the plates is the same as the potential difference between the battery terminals.

☐ The SI unit for capacitance is a coulomb/volt. This unit is called a farad and is abbreviated F. Microfarad (abbreviated μF and picofarad (abbreviated pF) capacitors are commonly used in electronic circuits.

26–3 Calculating the Capacitance

☐ To calculate capacitance, first imagine charge q is placed on one plate and charge $-q$ is placed on the other. Use Gauss' law to find an expression for the electric field in the region between the plates. The field at every point is proportional to q. Use $V = -\int \mathbf{E} \cdot d\mathbf{s}$ to compute potential difference V of the plates. V is also proportional to q. Finally, use $q = CV$ to find an expression for the capacitance.

☐ Suppose a capacitor consists of two parallel metal plates, each of area A, separated by a distance d. If charge q is paced on one and charge $-q$ is placed on the other, the electric field between the plates is $E = q/\epsilon_0 A$, the potential difference of the plates is $V = qd/\epsilon_0 A$, and the capacitance is $C = \epsilon_0 A/d$. The capacitance depends on the permittivity constant ϵ_0, the plate area A, and the plate separation d.

☐ Suppose a capacitor consists of two concentric cylindrical shells with length L, the inner one with radius a and the outer one with radius b. Charge q is placed on the inner cylinder and charge $-q$ is placed on the outer cylinder. The electric field between the cylinders, a distance r from the axis, is $E = q/2\pi\epsilon_0 Lr$, the potential difference of the cylinders is $V = (q/2\pi\epsilon_0 L)\ln(b/a)$, and the capacitance is $C = 2\pi\epsilon_0 L/\ln(b/a)$. C depends only on ϵ_0, a, b, and L.

☐ Suppose a capacitor consists of two concentric spherical shells, with radii a and b. Charge q is placed on the inner shell and charge $-q$ is placed on the outer shell. The electric field between the shells, a distance r from the sphere centers, is $E = q/4\pi\epsilon_0 r^2$, the potential difference of the shells is $V = q(b - a)/4\pi\epsilon_0 ab$, and the capacitance is $C = 4\pi\epsilon_0 ab/(b - a)$.

☐ Capacitance is also defined for a single conductor. The other plate is assumed to be infinitely far away. To find an expression for the capacitance, imagine charge q is on the conductor, find an expression for the electric field outside the conductor, calculate the potential V of the conductor relative to the potential at infinity, and use $q = CV$ to find the capacitance. For example, the capacitance of a spherical conductor of radius R is $C = 4\pi\epsilon_0 R$.

26–4 Capacitors in Parallel and in Series

☐ The potential difference is the same for all capacitors in a parallel connection; the charge is the same for all capacitors in a series connection. If neither the potential difference nor the charge is the same, then the capacitors do not form either a parallel or series combination.

☐ The capacitors in either a series or a parallel connection can be replaced by a single capacitor with capacitance C_{eq} such that the charge transferred is the same as the total charge transferred for the original combination when the potential difference is the same as the potential difference across the original combination. If the capacitances are C_1 and C_2, then the value of C_{eq} for a parallel combination is $C_{eq} = C_1 + C_2$ and the value of C_{eq} for a series combination is $C_{eq} = C_1 C_2/(C_1 + C_2)$. For a parallel combination, the equivalent capacitance is greater than the greatest capacitance in the combination; for a series combination, the equivalent capacitance is less than the smallest capacitance in the combination.

26–5   Storing Energy in an Electric Field

☐ As a capacitor is being charged, energy must be supplied by an external source, such as a battery, and energy is stored by the capacitor. You may think of the stored energy in either one of two ways: as the potential energy of the charge on the plates or as an energy associated with the electric field produced by the charges.

☐ Suppose that, at one stage in the charging process, the positive plate has charge q' and the potential difference across the plates is V'. If an additional infinitesimal charge dq' is taken from the negative plate and placed on the positive plate, the energy is increased by $dU = V' dq'$ or, what is the same, by $dU = (q'/C) dq'$, where C is the capacitance. This expression is integrated from 0 to the final charge q to obtain $U = q^2/2C$. It is also given by $U = CV^2/2$.

☐ The energy density (or energy per unit volume) in a parallel plate capacitor is given by

$$u = \tfrac{1}{2}\epsilon_0 E^2 \, ,$$

where E is the magnitude of the electric field between the plates. This expression is quite generally valid for any electric field, not only those in capacitors. Most fields are functions of position so a volume integral must be evaluated to calculate the energy required to produce the field: $U = \tfrac{1}{2}\epsilon_0 \int E^2 \, dV$.

☐ If, after charging a capacitor, the plates are connected by a conducting wire, then electrons will flow from the negative to the positive plate until both plates are neutral and the electric field vanishes. The stored energy is converted to kinetic energy of motion and eventually to internal energy in resistive elements of the circuit.

26-6 Capacitor with a Dielectric

☐ If the space between the plates of a capacitor is filled with insulating material, the capacitance is greater than if the space is a vacuum. If the space is completely filled, the capacitance is given by $C = \kappa C_0$, where C_0 is the capacitance of the unfilled capacitor and κ is the **dielectric constant** for the insulator. This last quantity is a property of the insulator, is unitless, and is always greater than unity.

☐ The energy stored in a capacitor with a dielectric is $U = \frac{1}{2}q^2/C = \frac{1}{2}CV^2$, where q is the magnitude of the charge on either plate and V is the potential difference. As a dielectric is inserted between the plates of a capacitor with the potential difference held constant (by a battery, say), both the charge on the positive plate and the stored energy increase. For the potential difference to remain the same, the battery must transfer charge as the dielectric is inserted. As a dielectric is inserted with the capacitor in isolation, so the charge cannot change, both the potential difference and the stored energy deceases. If the dielectric is inserted by an external agent, the difference in energy is associated with the work done by the agent.

26-7 Dielectrics: An Atomic View

☐ **Polarization** of the dielectric by the electric field causes an increase in capacitance. If the dielectric is composed of polar molecules, they rotate to align with the electric field between the plates. If the dielectric is composed of nonpolar molecules, dipoles are induced on the molecules by the field.

☐ Electric dipoles in a polarized dielectric produce an electric field that is directed opposite to the field produced by the charge on the conducting plates. Thus, the total field is weaker than the field produced by charge on the plates alone. It is, in fact, weaker by the factor κ. That is, if $\mathbf{E}_0$ is the field produced by the charge on the plates, then the total field is given by $E = E_0/\kappa$. If the electric field produced by charge on the plates is uniform, as it essentially is between the plates of a parallel plate capacitor, then the dipole field and the total field are also uniform.

☐ Since the electric field for a given charge on the plates is weaker if the capacitor is filled with a dielectric than if it is not, the potential difference is less when the dielectric is present. Since $q = CV$, this means the capacitance is larger when the dielectric is present.

26–8 Dielectrics and Gauss' Law

☐ For a parallel plate capacitor, the effect of a polarized dielectric is exactly the same as a uniform distribution of positive charge q' on the dielectric surface nearest the negative plate and a uniform distribution of negative charge $-q'$ on the dielectric surface nearest the positive plate. If the dielectric constant of the dielectric is κ and the charge on the positive plate is q, then $q' = (\kappa - 1)q/\kappa$. If the plates have area A, then Gauss' law can be used to show that the electric field due to the charge on the plates is $q/\epsilon_0 A$, the field due to the dipoles is $q'/\epsilon_0 A = (\kappa - 1)q/\kappa\epsilon_0 A$, and the total field is $E = (q/\epsilon_0 A) - [(\kappa - 1)q/\kappa\epsilon_0 A] = q/\kappa\epsilon_0 A$.

NOTES:

Chapter 26: Capacitance

Chapter 27
CURRENT AND RESISTANCE

Here you begin the study of electric current, the flow of charge. Pay close attention to the definitions of current and current density and understand how they depend on the concentration and speed of the charges. For most materials, a current is established and maintained only when an electric field is present. You will learn about the properties of materials that determine the magnitude and direction of the current for any given field.

Important Concepts

☐ electric current
☐ current density
☐ resistance
☐ resistor
☐ resistivity
☐ conductivity

☐ temperature coefficient
  of resistivity
☐ mean free time
☐ energy dissipated by a resistor
☐ semiconductor
☐ superconductor

27–1   Moving Charges and Electric Currents

☐ **Electric current** is the flow of charge. When a material object, such as a wire, contains a current, it is usually electrons that flow, not atomic nuclei. However, current from the sun consists of both electrons and protons and cosmic rays are chiefly highly energetic protons.

☐ Positive and negative charges moving in the same direction are currents in opposite directions. If the two currents have the same magnitude, the net current is zero. There is no net transport of charge. Thus, a neutral moving object consists of moving electrons and protons but does not constitute an electric current. The random motion of electrons in a piece of metal do not constitute a current because at any instant just as many are moving in any one direction as there are moving in the opposite direction. Again, there is no net transport of charge. When a battery is connected across the metal, an organized motion, opposite the electric field, is superimposed on the random motion of the electrons. Charge is now transported and there is a current.

27–2 Electric Current

☐ The current i through any cross-sectional area of a conducting wire is given by

$$i = \frac{dq}{dt},$$

where dq is the net charge that passes through the area in the time interval dt. Although current is a scalar, it is assigned a direction. Positive charge moving to the right and negative charge moving to the left are both currents to the right. For steady flow of charge through a conductor, the current is the same for all cross sections, no matter how the cross-sectional area differs along the wire, because equal charge passes every cross section in equal times.

☐ The SI unit for current is the coulomb/second. This unit is called an ampere and is abbreviated A.

27–3 Current Density

☐ Current is associated with an area, like the cross-sectional area of a conductor. On the other hand, **current density J** is a related quantity that can be associated with each *point* in a conductor. At any point, it is the current per unit area through an area at the point, oriented perpendicularly to the charge flow, in the limit as the area shrinks to the point. Current density is a vector. If positive charge is flowing, **J** is in the direction of the velocity; if negative charge is flowing, **J** is opposite the direction of the velocity.

☐ Consider any surface within a current-carrying conductor and let d**A** be an infinitesimal vector area, with direction perpendicular to the surface. If **J** is the current density, then the current i through the surface is given by the integral over the surface:

$$i = \int \mathbf{J} \cdot d\mathbf{A} \, .$$

The scalar product indicates that only the component of **J** normal to the surface contributes to the current through the surface. If the surface is a plane with area A and is perpendicular to the particle velocity and if the current density is uniform over the surface, then $i = JA$.

☐ For a collection of particles with uniform concentration n, the current density is given by the vector relationship

$$\mathbf{J} = en\mathbf{v}_d \, ,$$

where e is the charge on any one of the particles and v_d is the average velocity of the particles (the drift velocity). This expression can be derived by considering the number of particles per unit time that pass through an area perpendicular to their velocity. **J** and $\mathbf{v}_d$ are in opposite directions for negatively charged particles.

☐ The velocity of an electron can be considered to be the sum of two velocities, one of which changes direction often as the electron collides with atoms. When all electrons are taken into account, this velocity averages to zero and so it does not contribute to the current. The second component, the drift velocity, is much smaller in magnitude than the first. This is the velocity that enters the expression for the current density, not the total velocity.

☐ In an ordinary conductor, an electric field is required to produce a net drift and hence an electric current. For electrons, the direction of the drift velocity is opposite the direction of the electric field. This means the current is in the same direction as the field and it is directed from a region of high electric potential toward a region of lower electric potential.

☐ That an electric field can be maintained in a conductor does not contradict the statement made earlier that the electrostatic field in a conductor is zero. A conductor containing a current is not in electrostatic equilibrium.

27–4 Resistance and Resistivity

☐ **Resistance** is a measure of the current generated in a given conductor by a given potential difference. To determine resistance, a potential difference V is applied between two points in the material and the current i is measured. The resistance for that material and those points of application is defined by

$$R = \frac{V}{i}.$$

The resistance depends on where the current leads are attached, as well as on the material and its geometry. Resistance has an SI unit of volt/ampere, a unit that is called an ohm and is abbreviated Ω. In drawing an electrical circuit, an element whose function is to provide resistance, called a **resistor**, is indicated by the symbol —⋀⋀—.

☐ Resistance is intimately related to a property of the material called its **resistivity**. If, at some point in the material the electric field is $\mathbf{E}$ and the current density is $\mathbf{J}$, then the resistivity ρ at that point is given by $\mathbf{E} = \rho\mathbf{J}$. The resistivities of the materials we consider are the same at every point in the material and are the same for every orientation of the electric field. The SI unit for resistivity is $\Omega \cdot \mathrm{m}$.

☐ Typical semiconductors have resistivities that are greater than those of metals by factors of 10^5 to 10^{11} and insulators have resistivities that are greater than those of metals by factors of 10^{18} to 10^{24}. The **conductivity** σ of any substance is related to the resistivity by $\sigma = 1/\rho$.

☐ Let L be the length of a homogeneous wire with uniform cross section, A be its cross-sectional area, and ρ be its resistivity. If one end of the wire is held at potential 0 and the other is held at potential V, the electric field in the wire is given by $E = V/L$. If the current is i then, since the current density is uniform, it is given by $J = i/A$. Substitute $E = V/L$ and $J = i/A$ into $E = \rho J$ and solve for V/i ($= R$). The result is $R = \rho L/A$. The resistance depends on a property of the material (ρ) and on the geometry of the sample (L and A).

☐ The resistivity of a metal increases with increasing temperature, chiefly because electrons suffer more collisions per unit time at high temperatures than at low temperatures. The temperature dependence is characterized by a quantity α, called the **temperature coefficient of resistivity**, which is a measure of the deviation of the resistivity from its value at a reference temperature. Let ρ_0 be the resistivity at the reference temperature T_0 and let ρ be the resistivity at a nearby temperature T. Then, in terms of α,

$$\rho - \rho_0 = \rho_0 \alpha (T - T_0).$$

Temperatures are given in K or °C.

☐ Table 27–1 lists values for some materials. Notice that α is negative for semiconductors. For them, the resistivity decreases as the temperature increases because the concentration n of nearly free electrons increases rapidly with temperature. For metals near room temperature, n is essentially independent of temperature.

27–5 Ohm's Law

☐ Ohm's law describes an important characteristic of the resistance of certain samples, called ohmic samples (or devices). It is: For an ohmic device, the current is proportional to the potential difference. $V = iR$ defines the resistance R and so holds for every sample. For an ohmic sample, V is a *linear* function of i or, what is the same, R does not depend on V or i. In addition, for ohmic samples a reversal of the potential difference simply reverses the direction of the current without changing its magnitude. For some non-ohmic samples, such as a *pn* junction diode, a reversal of the potential difference not only changes the direction of the current but also its magnitude.

27–6 A Microscopic View of Ohm's Law

☐ For an ohmic substance, the resistivity is independent of the electric field and the current density is therefore proportional to the electric field. This implies that the drift velocity is also proportional to the field. Since an electric field accelerates charges, you should be surprised.

☐ To understand why Ohm's law is valid for some materials, consider the collection of free electrons in a conducting sample. These electrons are accelerated by the electric field applied to the sample and suffer collisions with atoms of the material. On average, the effect of a collision is to stop the drift of an electron so that, as far as drift is concerned, an electron starts from rest at a collision and is accelerated by the field until the next collision. The collision stops it and the process is repeated. The time between collisions, averaged over all electrons, is called the **mean free time** and is designated by τ. Because collisions occur at random times, τ is independent of the time; its value is the same no matter when the averaging is done.

☐ The magnitude of the acceleration is $a = qE/m$, where E is the magnitude of the electric field and m is the mass of an electron, so the average electron drift speed is given by $v_d = a\tau = e\tau E/m$. Although each electron accelerates between collisions, v_d does not vary with time if the field is constant. Also note that v_d is proportional to E if τ does not depend on E. The drift speed is such a small fraction of the electron's speed that the electric field does not appreciably alter its speed and, therefore, does not significantly change the mean time between collisions.

☐ The current density is given by $J = env_d = (e^2 n\tau/m)E$. This expression immediately gives $\rho = m/e^2 n\tau$ for the resistivity. Because τ is independent of E, the resistivity is also independent of E and the material obeys Ohm's law.

☐ A long mean free time leads to a small resistivity because an electron is accelerated for a long time before it is stopped by a collision. The drift velocity is great so the resistivity is small. Similarly, a short mean free time implies a small drift velocity and a large resistivity.

☐ Because the mean time between collisions is determined to a large extent by thermal vibrations of the atoms, it is temperature dependent. As the temperature increases, an electron suffers more collisions per unit time, so τ becomes shorter as the temperature increases.

27–7 Power in Electric Circuits

☐ If current i passes through a potential difference V, from high to low potential, the moving charges lose potential energy at a rate given by $P = dU/dt = iV$. For a resistor, the energy is transferred to atoms of the material in collisions. The result is an increase in the internal energy of the resistor and is usually accompanied by an increase in temperature. The phenomenon finds practical application in toasters and electrical heaters.

☐ Since $V = iR$ for a resistor, the expression for the rate of energy **dissipation** can be written as $P = i^2R$ or as $P = V^2/R$.

27–8 Semiconductors

☐ The resistivities of semiconductors are greater than the resistivity of a typical metal but less than the resistivity of a typical insulator. The concentration of nearly free electrons is intermediate between the concentration for a conductor and the concentration for an insulator. The nearly free electron concentration in a semiconductor can be adjusted by the addition of certain impurities.

☐ The resistivity of a semiconductor, unlike that of a metal, decreases with increasing temperature. Although the mean free time decreases with increasing temperature, like that of a metal, the number of electrons in the current increases dramatically as the temperature increases.

☐ The special properties of semiconductors make them useful in a wide variety of electronic devices.

27–9 Superconductors

☐ When the temperature of a superconductor is lowered below what is called its critical temperature, its resistivity becomes zero. Only certain materials become superconducting but they are of great technological value because they can carry large currents without heating. Electromagnets, motors, and magnetic field detectors make use of superconducting wires. Materials are being sought that turn superconducting at room temperatures and above, although none have been found as yet.

NOTES:

Chapter 28
CIRCUITS

The concept of emf is introduced in this chapter. Emf devices are used to maintain potential differences and to drive currents in electrical circuits. You will use this concept, along with those of electric potential, current, resistance, and capacitance, learned earlier, to solve both simple and complicated circuit problems. You will also learn about energy balance in electrical circuits.

Important Concepts

☐ emf device
☐ emf
☐ battery
☐ internal resistance
☐ energy in a circuit
☐ Kirchhoff loop rule
☐ Kirchhoff junction rule

☐ resistors in series
☐ resistors in parallel
☐ voltmeter
☐ ammeter
☐ charging a capacitor
☐ discharging a capacitor
☐ capacitive time constant

28–1 "Pumping" Charges

☐ A potential difference must be maintained for a current to exist in a resistor. An **emf device**, such as a battery, produces the required potential difference. It "pumps" charge around a circuit.

28–2 Work, Energy, and Emf

☐ An emf device performs two functions: it maintains a potential difference and it moves charge from one terminal to the other inside the device. An ideal emf device contains only a seat of emf; a real emf device, such as a **battery**, also contains **internal resistance**.

☐ In electrical circuits, an ideal emf device is indicated by the symbol ⊶⊣⊢. The arrow with the small circle at its tail points from the negative terminal toward the positive terminal. An emf tends to drive current in the direction of the arrow but, for any circuit, the actual direction of the current through an emf device depends on other elements in the circuit.

☐ An **emf**, denoted by $\mathcal{E}$, is defined in terms of the work it does as it transports charge. If it does work dW on charge dq, then $\mathcal{E} = dW/dq$. The rate at which an emf does work is $dW/dt = (dq/dt)\mathcal{E} = i\mathcal{E}$. Emf has an SI unit of volt.

☐ An emf device does positive work on the charge passing through it if the current is from the negative to the positive terminal and negative work if the current is in the other direction. This is true whether the current consists of positive or negative charges.

☐ Positive work done by an emf device results in a decrease in the store of energy of the device, chemical energy in the case of a battery. Negative work results in an increase in the store of energy of the device. If the device is a battery, then in the first case it is *discharging* and in the second it is *charging*.

☐ The potential difference across an ideal emf device with emf $\mathcal{E}$ is $\mathcal{E}$, with the positive terminal being at the higher potential. This statement is true no matter what the direction of the current through the device.

28–3 Calculating the Current in a Single-Loop Circuit

☐ Charge flows if a resistor is connected across an emf device. In steady state, the rate with which the emf does work on the charge must equal the rate with which energy is dissipated in the resistor: $i\mathcal{E} = i^2 R$. Thus, the current is $i = \mathcal{E}/R$.

☐ Circuits in steady state obey the **Kirchhoff loop rule**: The algebraic sum of the changes in potential encountered in a complete traversal of the circuit must be zero. To apply the rule, pick a direction for the current in the loop, then traverse the circuit. If you cross an emf from the negative to the positive terminal, the change in potential is $+\mathcal{E}$. If you traverse it in the other direction, the change is $-\mathcal{E}$. If you traverse a resistor in the direction of the current arrow, the change in potential is $-iR$; if you traverse it in the other direction the change is $+iR$. These expressions are valid even if you picked the wrong direction for the current arrow; i then has a negative value.

☐ For a single-loop circuit consisting of an ideal emf device with emf $\mathcal{E}$ and a resistor with resistance R, the loop rule gives $\mathcal{E} - iR = 0$, in agreement with the energy equation.

28–4 Other Single-Loop Circuits

☐ A real battery contains an internal resistance in addition to a seat of emf. Although the internal resistance is distributed inside the battery, you may think of it as being in series with an ideal emf.

☐ Suppose a single-loop circuit consists of a battery with emf $\mathcal{E}$ and internal resistance r, connected to an external resistance R. Then, the loop rule gives $\mathcal{E} - ir - iR = 0$ and the current in the circuit is $i = \mathcal{E}/(r + R)$. Study Sample Problems 28–1 and 28–2 of the text for an example of a circuit with two batteries. Note that the battery with the larger emf is discharging and determines the direction of the current while the battery with the smaller emf is charging.

28–5 Potential Differences

☐ Once the current in a circuit is determined, its value can be used to calculate the potential difference across any of the circuit elements. The potential difference across a resistor, for example, is iR. The potential difference across two or more circuit elements in series is the sum of the individual potential differences.

28–6 Multiloop Circuits

☐ To analyze multiloop circuits, the **Kirchhoff junction rule** is needed in addition to the loop rule. The rule is: At any junction, the sum of the currents entering the junction is equal to the sum of the currents leaving. If, for example, four wires join at a junction, with i_1 and i_2 going into the junction and i_3 and i_4 coming out, then $i_1 + i_2 = i_3 + i_4$. The form of the junction equation does not depend on the directions of the actual currents, only on the directions of the current arrows.

☐ A junction is a point where three or more wires are joined. A branch is a portion of a circuit with junctions at its ends and none between. You must place a current arrow in each branch and label the arrows in different branches with different current symbols. In general, for any circuit with N junctions there will be $N - 1$ independent junction equations. The total number of independent equations equals the number of branches so the number of independent loop equations you will use equals the number of branches minus the number of junction equations. The loop and junction equations can be solved simultaneously for the currents in all the branches.

☐ The potential difference is the same for all the resistors in a parallel combination and the current is the same for all resistors in a series combination. If neither the potential difference nor the current is the same, then the resistors do not form either a parallel or series combination.

☐ For both series and parallel combinations the resistors can be replaced by a single resistor with resistance R_{eq}, such that the current is the same as the total current into the original combination when the potential difference is the same as the potential difference across the original combination. For a series combination of N resistors,

$$R_{eq} = \sum_{i=1}^{N} R_i$$

and for a parallel combination

$$\frac{1}{R_{\text{eq}}} = \sum_{i=1}^{N} \frac{1}{R_i}.$$

28–7   The Ammeter and the Voltmeter

☐   An **ammeter** measures the current in a branch of a circuit. The circuit is broken and the meter is inserted, so the current in the branch goes through the meter. An ammeter has a resistance and this alters the current it is measuring. To minimize this effect, the resistance of the ammeter should be very small.

☐   A **voltmeter** measures the potential difference between two points in a circuit. The connection is made so the meter is in *parallel* with a portion of the circuit. Voltmeters draw current and thereby change the value of the potential difference they are measuring. To minimize this effect, the resistance of the meter should be very large.

28–8   *RC* Circuits

☐   A capacitor is charged by connecting it in series with a resistor and an emf to form a single-loop circuit. The loop rule gives $\mathcal{E} - iR - (q/C) = 0$. Here the direction of the current arrow was chosen to be in the direction of the emf and q represents the charge on the capacitor plate to which the current arrow points. This means the current and charge are related by $i = \mathrm{d}q/\mathrm{d}t$ and the equation satisfied by the charge is

$$\mathcal{E} - R\frac{\mathrm{d}q}{\mathrm{d}t} - \frac{q}{C} = 0.$$

☐   If the charging process is started at time $t = 0$ (by throwing a switch then, for example), the charge is given by

$$q(t) = C\mathcal{E}\left(1 - e^{-t/RC}\right)$$

and the current is given by

$$i(t) = \frac{\mathrm{d}q}{\mathrm{d}t} = \left(\frac{\mathcal{E}}{R}\right) e^{-t/RC} .$$

The solution obeys the initial condition $q(0) = 0$. These functions are graphed in Fig. 28–14.

☐ According to this result, the current just after the switch is closed, at $t = 0$, is given by $i(0) = \mathcal{E}/R$, just as if there were no capacitor. The results also predict that after a long time the charge on the capacitor is $q(\infty) = C\mathcal{E}$ and the current is $i(\infty) = 0$. The capacitor is then fully charged.

☐ The quantity $\tau = RC$ is called the **capacitive time constant** of the circuit. It has units of seconds and controls the time for the charge on the capacitor to reach any given value. If τ is made longer (by increasing C or R), more time is taken for the capacitor to reach any given fraction of its full charge $C\mathcal{E}$.

☐ A capacitor C, initially with charge q_0, is discharged by connecting it to a resistor R. The loop equation is

$$R\frac{\mathrm{d}q}{\mathrm{d}t} + \frac{q}{C} = 0$$

and its solution is

$$q(t) = q_0 \, e^{-t/RC} .$$

After a long time, q becomes zero. The current as a function of time is

$$i(t) = \frac{\mathrm{d}q}{\mathrm{d}t} = -\frac{q_0}{RC} \, e^{-t/RC} .$$

Initially, the current is $i(0) = q_0/RC$; after a long time, it is zero.

NOTES:

Chapter 29
MAGNETIC FIELDS

Charged particles, when moving, exert magnetic as well as electric forces on each other. Magnetic fields are used to describe magnetic forces and, although the geometry is a little more complicated, the idea is much the same as the idea of using electric fields to describe electric forces. Here you will learn about the force exerted by a magnetic field on a moving charge and on a wire carrying an electric current. In each case, pay particular attention to what determines the magnitude and direction of the force.

Important Concepts

☐ magnetic field
☐ magnetic force on a charge
☐ Hall effect
☐ orbit of a charge
    in a magnetic field

☐ magnetic force on a
    current-carrying wire
☐ magnetic torque
    on a current loop
☐ magnetic dipole
☐ magnetic dipole moment

29–1   The Magnetic Field

☐ A moving charge creates a **magnetic field** in all of space and this magnetic field exerts a force on any other moving charge. Charges at rest do not produce magnetic fields nor do magnetic fields exert forces on them.

29–2   The Definition of B

☐ The force exerted by the magnetic field **B** on a charge q moving with velocity **v** is given by the vector product

$$\mathbf{F}_B = q\mathbf{v} \times \mathbf{B} .$$

The magnitude of $\mathbf{v} \times \mathbf{B}$ is given by $vB \sin \phi$, where ϕ is the angle between $\mathbf{v}$ and $\mathbf{B}$ when they are drawn with their tails at the same point. The direction of $\mathbf{v} \times \mathbf{B}$ is determined by the right-hand rule: curl the fingers of the right hand so they rotate $\mathbf{v}$ into $\mathbf{B}$ around the point where their tails meet, through the angle ϕ. The magnetic force is zero if $\mathbf{v}$ is parallel to $\mathbf{B}$. It is a maximum if $\mathbf{v}$ is perpendicular to $\mathbf{B}$.

☐ A magnetic force is perpendicular to the magnetic field and to the velocity of the charge. Because the force is perpendicular to the velocity, a magnetic field cannot change the speed or kinetic energy of a charge. It can, however, change the direction of motion.

☐ The sign of the charge is important for determining the direction of the force. A positive charge and a negative charge moving with the same velocity in the same magnetic field experience magnetic forces in opposite directions. The force on a positive charge is in the direction of $\mathbf{v} \times \mathbf{B}$ while the force on a negative charge is in the opposite direction.

☐ The magnetic field is defined in terms of the force on a moving positive test charge. First, the electric force is found by measuring the force on the test charge when it is at rest. This force is subtracted vectorially from the force on the test charge when it is moving in order to find the magnetic force. Second, the direction of the magnetic field is found by causing the test charge to move in various directions and seeking the direction for which the magnetic force is zero. Lastly, the test charge is given a velocity perpendicular to the field and the magnetic force on it is found. If the charge is q_0, its speed is v, and the magnitude of the force on it is F_B, then the magnitude of the magnetic field is given by $B = F_B/q_0 v$.

☐ The SI unit for a magnetic field is the tesla and is abbreviated T: $1\,\text{T} = 1\,\text{N/A} \cdot \text{m}$. Another unit in common use is the gauss: $1\,\text{T} = 10^4$ gauss.

Chapter 29: Magnetic Fields

☐ Magnetic field lines are drawn so that at any point the field is tangent to the field line through that point and so that the magnitude of the field is proportional to the number of lines through a small area perpendicular to the field.

☐ Magnetic field lines form closed loops. For example, they continue through the interior of a magnet. The end of a magnet from which they emerge is called a north pole; the end they enter is called a south pole. If a bar magnet is free to rotate in Earth's magnetic field, its north pole tends to point roughly toward the north geographic pole.

29–3 Crossed Fields: Discovery of the Electron

☐ If both a magnetic field **B** and an electric field **E** act simultaneously on a charge q moving with velocity **v**, the total force on the charge is given by

$$\mathbf{F} = q\mathbf{E} + q\mathbf{v} \times \mathbf{B} .$$

An electric force can be used to balance a magnetic force on a charge, but the magnitude and direction of the electric field depend on the velocity of the charge. If a charge is in a magnetic field **B** and has velocity **v**, then the total force is zero if $\mathbf{E} = -\mathbf{v} \times \mathbf{B}$. This result does not depend on either the sign or magnitude of the charge. The balancing electric field is perpendicular to both the magnetic field and the velocity of the charge. An important special case occurs if the velocity of the charge is perpendicular to the magnetic field. Then, the magnitude of the balancing electric field is $E = vB$.

☐ Crossed electric and magnetic fields can be used to measure the mass-to-charge ratio m/q of an electron (or other charged particle). A beam of electrons, emitted from a hot filament, is formed by accelerating them in an electric field and passing them through a slit. They then enter a region of crossed electric and magnetic fields. The beam hits a fluorescent screen and produces a spot where it hits. The steps of the measurement are:

1. The fields are turned off and the position of the spot is noted.

2. Only the electric field **E** is turned on and the deflection y of the beam at the edge of the deflecting plates is computed from the deflection of the spot on the screen. It is given by $y = qEL^2/2mv^2$, where L is the length of the plates and v is the speed of the particles.

3. The magnetic field is now adjusted until the spot returns to its original position. Then, the magnetic and electric forces cancel and the speed is $v = E/B$. Substitute this expression into the equation for y and solve for m/q. The result is $m/q = B^2L^2/2yE$. All of the quantities on the right side can be determined experimentally.

29–4 Crossed Fields: The Hall Effect

☐ A Hall effect experiment can be used to obtain the sign and concentration of charges in a current. A magnetic field, perpendicular to the current, pushes charges to one side of the sample. These charges create a transverse electric field and when there are enough of them, the electric force on charges in the current cancels the magnetic force. Then no more charge collects along the side. The direction of the transverse electric field and, hence, the sign of the charges can be found with a voltmeter. The concentration of particles in the current is given by $n = iB/qV\ell$, where i is the current, B is the magnetic field, V is the transverse potential difference, and ℓ is the thickness of the sample.

29–5 A Circulating Charged Particle

☐ Since a magnetic force is always perpendicular to the velocity of the charge on which it acts, it can be used to hold a charge in a circular orbit. All that is required is to fire the charge perpendicularly into a uniform field. Suppose a region of space contains a uniform magnetic field with magnitude B and a particle with charge of magnitude q is given a velocity **v** perpendicular to the field. Equate the magnitude of the magnetic force (qvB) to the product of the mass and acceleration (mv^2/r) and solve for the radius r of the orbit. The result is $r = mv/qB$.

☐ The time taken by the charge to go once around its orbit does not depend on its speed because the radius of the orbit and, hence, the distance traveled are both proportional to the speed. Since the circumference of the orbit is $2\pi R$, the period is $T = 2\pi r/v = 2\pi m/qB$. The number of times the charge goes around per unit time, or the frequency of the motion, is given by $f = 1/T = qB/2\pi m$. The angular frequency is given by $\omega = 2\pi f = qB/m$.

☐ If the particle velocity is not perpendicular to the field, but has both parallel and perpendicular components, the trajectory is a helix. The radius is determined by the velocity component in the plane perpendicular to the field and the pitch is determined by the velocity component parallel to the field.

29–6   Cyclotrons and Synchrotrons

☐ A cyclotron is diagramed in Fig. 29–15. A uniform magnetic field is everywhere perpendicular to the dees and an electric field is in the gap between them. A charged particle enters the cyclotron near the center. The magnetic field causes the charge to travel in a circular arc. The electric field causes the speed of the charge to increase and when it does, the charge moves to an orbit of larger radius. The electric field must reverse direction twice each period if it is to increase the speed of the charge during every traversal of the gap. This is fairly easy to do since the time between reversals does not change as long as the particle speed is significantly less than the speed of light.

☐ At speeds near the speed of light, the period does depend on the particle speed and the interval between reversals of the electric field must be decreased as the charge speeds up. Synchrotrons are accelerators that change the frequency of the electric field so its reversals remain in step with the circulating charges. The magnetic field is also varied so the radius of the orbit remains the same.

29–7 Magnetic Force on a Current-Carrying Wire

- ☐ A current is a collection of moving charges and so experiences a force when a magnetic field is applied. If the current is in a wire, the force is transmitted to the wire itself.

- ☐ To calculate the force on a wire carrying current i, divide the wire into infinitesimal segments, calculate the force on each segment, then vectorially sum the forces. Let dL be the length of an infinitesimal segment of wire with cross-sectional area A. If n is the concentration of charge carriers and each carrier has charge q, then the total charge in the segment is given by $qnA\,dL$. The magnetic force on the segment is the product of this and the force on a single charge. That is, $d\mathbf{F}_B = qnA\mathbf{v}_d \times \mathbf{B}\,dL$, where $\mathbf{v}_d$ is the drift velocity. Notice that the combination $qnAv_d$ appears in the expression for the magnitude of the force. This is the current i. If we take the vector $d\mathbf{L}$ to be in the direction of $q\mathbf{v}_d$, then the force is given by $d\mathbf{F}_B = i\,d\mathbf{L} \times \mathbf{B}$.

- ☐ For a finite wire, the resultant force is given by the integral

$$\mathbf{F}_B = i \int d\mathbf{L} \times \mathbf{B}.$$

For a straight wire of length L in a uniform field $\mathbf{B}$, the magnitude of the force is $F_B = i\mathbf{L} \times \mathbf{B}$. Since $\int d\mathbf{L} = 0$ for a closed loop, the magnetic force on a closed loop of wire in a uniform magnetic field is zero.

29–8 Torque on a Current Loop

- ☐ Although a uniform magnetic field does not exert a net force on a current-carrying loop, it may exert a torque: the center of mass of the loop does not accelerate in a uniform field but if the magnetic torque is not balanced, the loop has an angular acceleration about the center of mass.

☐ Consider a rectangular loop of wire carrying current i, in a uniform magnetic field $\mathbf{B}$. If the sides of the loop have lengths a and b, then the magnetic torque (about any axis of rotation) is given by $\tau = iab\sin\theta$, where θ is the angle between the magnetic field and the normal to the plane of the loop. If the loop has N turns of wire, the torque is $\tau = (NiA)B\sin\theta$, where A ($= ab$) is the area of the loop.

☐ Let $\mathbf{n}$ be a unit vector that is normal to the loop, in the direction of the thumb when the fingers of the right hand curl around the loop in the direction of the current. Then, the magnetic torque is in the direction of $\mathbf{n} \times \mathbf{B}$.

☐ The torque has maximum magnitude when $\theta = 90°$; the normal to the loop is perpendicular to the field and the loop itself is parallel to the field. The torque is zero when $\theta = 0$; the normal to the loop is in the direction of the field and the loop itself is perpendicular to the field. The torque exerted by the magnetic field tends to rotate the normal toward the direction of the field.

29–9   The Magnetic Dipole

☐ The torque exerted by a uniform field on any planar current loop is easily expressed in terms of the **magnetic dipole moment** $\boldsymbol{\mu}$ of the loop. For a loop with area A, having N turns and carrying current i, the magnitude of the dipole moment is given by $\mu = NiA$. The direction of the dipole moment is determined by the right-hand rule: Curl the fingers of the right hand around the loop in the direction of the current. Then, the thumb points in the direction of the dipole moment.

☐ The torque exerted by a field $\mathbf{B}$ on a loop with magnetic dipole moment $\boldsymbol{\mu}$ is given by $\boldsymbol{\tau} = \boldsymbol{\mu} \times \mathbf{B}$. Its magnitude is given by $\tau = \mu B\sin\theta$, where θ is the angle between $\boldsymbol{\mu}$ and $\mathbf{B}$ when they are drawn with their tails at the same point. The direction of the torque is given by the right-hand rule for vector products.

☐ Although magnetic fields are not conservative and a potential energy cannot be associated with a charge in a magnetic field, a potential energy can be associated with a magnetic dipole μ in a magnetic field $\mathbf{B}$. It is given by $U = -\mu \cdot \mathbf{B}$. The potential energy is a minimum when the dipole moment is aligned with the magnetic field. It is a maximum when the dipole moment is antialigned with the field. In both these cases, the plane of the loop is perpendicular to the field.

☐ Suppose a dipole moment initially makes the angle θ_i with the magnetic field, but then rotates so it makes the angle θ_f. During this rotation, the work done on the dipole by the field is given by $W = -(U_f - U_i) = \mu B(\cos \theta_f - \cos \theta_i)$.

NOTES:

Chapter 30
MAGNETIC FIELDS DUE TO CURRENTS

Learn to calculate the magnetic field produced by a current using two techniques, one based on the Biot-Savart law and the other based on Ampere's law. When using the first, you will sum the fields produced by infinitesimal current elements. Carefully note how the directions of the current and the position vector from a current element to the field point influence the direction of the field. Ampere's law relates the integral of the tangential component of the field around a closed path to the net current through the path. Pay attention to the role played by symmetry when you use this law to find the field.

Important Concepts

☐ Biot-Savart law

☐ permeability constant

☐ magnetic field of a
   long straight wire

☐ magnetic field of an arc

☐ Ampere's law

☐ solenoid

☐ toroid

☐ magnetic field of a dipole

31–1   Calculating the Magnetic Field Due to a Current

☐ The **Biot-Savart law** is a prescription for calculating the magnetic field of a current. To calculate the magnetic field produced at any point P by a wire carrying a current i, first choose an infinitesimal element d**s** of the wire, in the same direction as the current. Draw the displacement vector **r** from the selected current element to P. The field produced at P by the current element is given by

$$d\mathbf{B} = \left(\frac{\mu_0}{4\pi}\right) \frac{i\,d\mathbf{s} \times \mathbf{r}}{r^3}.$$

The field of the element is perpendicular to both d**s** and **r**. To find the total field at P, sum (integrate) the contributions from all elements of the wire.

☐ The constant μ_0 is called the **permeability constant** and its value is exactly $4\pi \times 10^{-7}\,\text{T} \cdot \text{m/A}$. Do not confuse the symbol with that for the magnitude of a magnetic dipole moment (μ without a subscript).

☐ In Section 30–1 of the text, the Biot-Savart law is used to find an expression for the magnitude of the magnetic field produced by a long straight wire carrying current i. Each infinitesimal element of the wire produces a field in the same direction so the magnitude of the total field is the sum of the magnitudes of the fields produced by all the elements. Go over the calculation carefully. The result is $B = \mu_0 i/2\pi R$, where R is the perpendicular distance from the wire to the field point.

☐ The magnetic field lines around a long straight wire are circles centered on the wire. The field has the same magnitude at all points on any given circle. The direction is determined by a right-hand rule: If the thumb of the right hand points in the direction of the current, then the fingers curl around the wire in the direction of the magnetic field lines.

☐ You should also know how to compute the magnetic field produced at its center by a circular arc of current. Consider the wire shown below. It has radius R, subtends the angle ϕ, and carries current i. Other wires, not shown, carry the current into and out of the arc. Since the values of i and R are the same for all segments and the fields at C due to all segments are in the same direction (into the page), the magnitude of the total field at C is given by $B = (\mu_0/4\pi) \int (i/R^2)\, ds = \mu_0 is/4\pi R^2 = \mu_0 i\phi/4\pi R$, where s is the length of the arc. The relationship $s = R\phi$, with ϕ in radians, was used. For a semicircle, $\phi = \pi$ and $B = \mu_0 i/4R$ and for a complete circle, $\phi = 2\pi$ and $B = \mu_0 i/2R$.

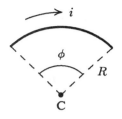

30–2 Two Parallel Currents

□   The expression for the field of a long straight wire can be
    combined with the expression for the magnetic force on a
    wire, developed in the last chapter, to find an expression
    for the force exerted by two parallel wires on each other.
    The magnetic field produced by one wire at a point on the
    other wire is perpendicular to the second wire. If the wires
    are separated by a distance d and carry currents i_a and i_b,
    then the magnitude of the force per unit length of one on
    the other is given by $F/L = \mu_0 i_a i_b/2\pi d$. If the currents are
    in the same direction, the wires attract each other; if the
    currents are in opposite directions, they repel each other.
    The forces of the wires on each other obey Newton's third
    law: they are equal in magnitude and opposite in direction.

30–3 Ampere's Law

□   Ampere's law tells us that the integral of the tangential
    component of the magnetic field around any closed path is
    equal to the product of μ_0 and the net current that pierces
    the loop. Mathematically, for any closed path,

$$\oint \mathbf{B} \cdot d\mathbf{s} = \mu_0 i,$$

    where $d\mathbf{s}$ is an infinitesimal displacement vector, tangent
    to the path. The integral on the left side is a path integral
    around a *closed* path, called an Amperian path. The cur-
    rent i on the right side is the net current through a surface

that is bounded by the path. For example, if the path is formed by the edges of a page, then i is the net current through the page. The Amperian path need not be the boundary of any physical surface and, in fact, may be purely imaginary. The surface need not be a plane.

☐ To apply Ampere's law, choose a direction (clockwise or counterclockwise) to be used in evaluating the integral on the left side. The choice is immaterial but it must be made since it determines the direction of ds. If the tangential component of the field is in the direction of ds, then the integral is positive; if it is in the opposite direction, the integral is negative. Now, curl the fingers of your right hand around the loop in the direction chosen. Your thumb will point in the direction of positive current. Examine each current through the surface and algebraically sum them. If a current arrow is in the direction your thumb pointed, it enters the sum with a positive sign; if it is in the opposite direction it enters with a negative sign.

☐ A current outside the path produces a magnetic field at every point on the path but the integral $\oint \mathbf{B} \cdot d\mathbf{s}$ of the field it produces vanishes, so it is not included on the right side of the equation.

☐ Only the tangential component of the magnetic field enters Ampere's law. When the law is used to calculate the magnetic field of a current distribution, the path is taken, if possible, to be along a field line. Then, the integral reduces to $\int B \, ds$. If, in addition, the magnitude of the field is constant along the path, then the integral is Bs, where s is the total length of the path.

☐ Ampere's law can be used to find an expression for the magnetic field produced by a long straight wire carrying current i. The Amperian path used is a circle in a plane perpendicular to the wire and centered at the wire. Since the magnetic field is tangent to the circle and has the same magnitude at all points around the circle, the integral $\oint \mathbf{B} \cdot d\mathbf{s}$ is $2\pi r B$, where r is the radius of the circle. Ampere's law gives $2\pi r B = \mu_0 i$, so $B = \mu_0 i/2\pi r$, as before.

☐ Ampere's law can also be used to find the magnetic field *inside* a long straight wire. Suppose a cylindrical wire of radius R carries a current that is uniformly distributed over its cross section. Take the Amperian path to be a circle of radius r (with $r < R$). The magnetic field is tangent to the circle and has constant magnitude around the circle, so $\oint \mathbf{B} \cdot d\mathbf{s} = 2\pi r B$. Not all the current in the wire goes through the Amperian circle. In fact, the fraction through the Amperian circle is the ratio of the circle area to the wire area: r^2/R^2. Thus, the right side of the Ampere's law equation is $\mu_0 i r^2/R^2$. Ampere's law gives $2\pi r B = \mu_0 i r^2/R^2$, so $B = \mu_0 i r/2\pi R^2$.

☐ The magnetic field at the center of the wire is zero. It increases linearly with distance from the wire center until it reaches its maximum value at the surface of the wire ($r = R$). This value is $B = \mu_0 i/2\pi R$.

30–4 Solenoids and Toroids

☐ Ampere's law can be applied to a **solenoid**, a cylinder tightly wrapped with a thin wire. For an ideal solenoid (long and tightly wrapped), the magnetic field outside is negligible and the field inside is uniform and is parallel to the axis.

☐ As the Amperian path, take a rectangle with one side, of length h, inside the solenoid and parallel to its axis, and the opposite side outside the solenoid. The first of these sides contributes Bh to the left side of the Ampere's law equation and the other three sides of the rectangle contribute zero. Thus, $\oint \mathbf{B} \cdot d\mathbf{s} = Bh$. If the solenoid has n turns of wire per unit length, then the number of turns that pass through the surface bounded by the Amperian path is nh. Each carries current i so the right side of the Ampere's law equation is $\mu_0 n h i$. Thus, $Bh = \mu_0 n h i$, so $B = \mu_0 n i$.

☐ Ampere's law can also be applied to a toroid, with a core shaped like a doughnut and wrapped with a wire, like a solenoid bent so its ends join. The magnetic field is confined to the interior of the core and the field lines are concentric circles centered at the center of the hole. The Amperian path is a circle of radius r, inside the core and centered at the center of the hole. The integral $\oint \mathbf{B} \cdot d\mathbf{s}$ is $2\pi r B$ and if there are N turns of wire, each carrying current i, the total current through the surface bounded by the path is Ni. Thus, the Ampere's law equation is $2\pi r B = \mu_0 Ni$ and $B = \mu_0 Ni/2\pi r$. The field is zero inside the hole and outside the toroid.

30–5 A Current-Carrying Coil as a Magnetic Dipole

☐ The Biot-Savart law can be used to show that the magnetic field produced by a circular loop of wire of radius R and carrying current i, at a point on its axis of symmetry a distance z from its center, is

$$B(z) = \frac{\mu_0 i R^2}{2(R^2 + z^2)^{3/2}} .$$

For points far away from the loop $(z \gg R)$, the expression becomes $B(z) = \mu_0 i R^2/2z^3$, which can be written in terms of the dipole moment of the loop: $B(z) = \mu_0 \mu/2\pi z^3$. $\mathbf{B}$ is in the direction of μ for points above the loop (positive z) and in the opposite direction for points below the loop (z negative). This expression is valid for the magnetic field of *any* plane loop, regardless of its shape, for points far away along the axis defined by the direction of the dipole moment.

NOTES:

Chapter 30: Magnetic Fields Due to Currents

Chapter 31
INDUCTION AND INDUCTANCE

A changing magnetic flux through any area induces an emf around the boundary and, if the boundary is conducting, also induces a current. Faraday's law gives the relationship between the emf and the rate of change of the flux. An emf can be generated by changing the magnetic field or by moving a physical object through a magnetic field. Concentrate on how to calculate the magnetic flux and the emf in each case. A non-conservative electric field is induced by a changing magnetic field and this field is responsible for the emf when the magnetic field is changing. You will also learn about inductive circuit elements, which produce emfs via Faraday's law when their currents are changing and which store energy in magnetic fields. Pay attention to the definition of inductance and learn to calculate its value for solenoids and toroids. Learn about the influence of inductance on the current in a circuit. Also learn how to calculate the energy stored in an inductor.

Important Concepts

☐ magnetic flux   ☐ flux linkage
☐ Faraday's law   ☐ inductor
☐ induced emf   ☐ RL circuit
☐ induced current   ☐ magnetic energy
☐ Lenz's law   ☐ magnetic energy density
☐ inductance   ☐ mutual inductance

31–1 Two Symmetric Situations

☐ When a current-carrying loop of wire is placed in a magnetic field, the field exerts a torque on the loop and the loop rotates. The reverse also happens. If the loop initially has no current but is then rotated in a magnetic field, current is induced in it.

31–2 Two Experiments

- [ ] If a magnet is moved toward a closed loop of wire, current is induced in the wire. When the magnet stops moving, the current stops. As the magnet is withdrawn, current is again induced. The direction of the current is reversed when the direction of motion of the magnet is reversed.

- [ ] If two loops of wire are close to each other and a changing current is produced in one, then current is induced in the second loop.

- [ ] In each of these cases, a changing magnetic field through a closed loop induces an emf around the loop and this emf generates a current. The emf is induced only when the magnetic field through the loop is changing or else when the loop is moving in a magnetic field.

31–3 Faraday's Law of Induction

- [ ] The **magnetic flux** through a surface is defined by the integral

$$\Phi_B = \int \mathbf{B} \cdot d\mathbf{A}$$

over the surface. Here d**A** is an infinitesimal vector area, normal to the surface. If the field is uniform over the surface, then $\Phi_B = BA\cos\theta$, where θ is the angle between the field and the normal to the surface.

- [ ] The magnetic flux through any area is proportional to the number of magnetic field lines through that area. Recall that the number of field lines through a small area perpendicular to the field is proportional to the magnitude of the field and that the number of lines through *any* small area is proportional to the component B_n along a normal to the area.

- [ ] The SI unit of magnetic flux is weber and is abbreviated Wb.

☐ Faraday's law tells us that whenever the flux through any area is changing with time, then an emf is generated around the boundary of that area. Symbolically, the law is

$$\mathcal{E} = -\frac{d\Phi_B}{dt},$$

where $\mathcal{E}$ is the induced emf. The induced emf is distributed around the boundary and is not localized as is the emf of a battery. Its direction is specified as clockwise or counterclockwise, for example. The negative sign that appears in Faraday's law is closely related to the direction of the emf.

☐ The direction of $d\mathbf{A}$ is ambiguous since two directions are normal to any surface. Φ_B has the same magnitude for both choices but is positive for one and negative for the other. You may choose either but you may find thinking about Faraday's law a bit easier if you pick the one for which the angle between $\mathbf{B}$ and $d\mathbf{A}$ is less than 90° and Φ_B is positive. If you point the thumb of your right hand in the direction chosen for $d\mathbf{A}$, then your fingers will curl around the boundary in the direction of positive emf. If $\mathcal{E}$, calculated using Faraday's law including the minus sign, turns out to be positive, then the emf is in the direction of your fingers. If it turns out to be negative, then it is in the opposite direction. If the boundary of the region is conducting and no other emf's are present, the induced current will be in the direction of the induced emf.

☐ Suppose a coil of wire consists of N identical turns, like a solenoid, with the same magnetic field through each turn. If Φ_B is the flux through each turn, then the total emf induced around the coil is given by $\mathcal{E} = -N\,d\Phi_B/dt$.

☐ The flux through the interior of a loop can be changed by changing the magnitude or direction of the magnetic field, by changing the area of the loop, or by changing the orientation of the loop. No matter how the emf is generated, it represents work done on a unit test charge as the test charge is carried around the loop.

31–4 Lenz's Law

☐ Lenz's law provides another way to determine the direction of an induced emf. Imagine that the boundary of an area is a conducting wire and that an externally produced magnetic field penetrates the area. When the field changes, a current is induced in the wire and the current also produces a magnetic field. According to Lenz's law, the sign of the flux of the induced current is the same as that of the externally produced flux if that flux is decreasing and is opposite that of the externally produced flux if that flux is increasing. In the first case, the magnetic field of the induced current is roughly in the same direction as the externally produced field; in the second case, it is roughly in the opposite direction.

☐ Once you have determined the direction of the magnetic field produced by the induced current, you can determine the direction of the current itself and, hence, that of the emf. Very near any segment of the loop, the field is quite similar to the field of a long straight wire; the lines are nearly circles around the segment. Use the right-hand rule explained in the last chapter: curl your fingers around the segment so they point in the direction of the field in the interior of the loop. That is, they should curl upward through the loop if the field is upward through the loop and they should curl downward if the field is downward. Your thumb will then point along the segment in the direction of the current. This is also the direction of the induced emf.

31–5 Induction and Energy Transfers

☐ An emf is also generated if all or part of the loop moves in a manner that changes the flux through it. If a loop is moved through a uniform field, the flux through it does not change and no emf is generated. If, however, the field is not uniform, then the flux changes as the loop moves and an emf is generated. Find an expression for the flux as a function of the position of the loop, then differentiate it with respect to time to obtain $d\Phi_B/dt$. The emf clearly depends on the velocity of the loop. Either Lenz's law or the sign convention associated with Faraday's law can be used to find the direction of the emf.

☐ Suppose a closed loop is pulled with constant speed out of the region of a uniform magnetic field, perpendicular to the loop, as shown below. At any instant, x is the length of the loop within the region of the field and the flux through the loop is $\Phi = BLx$. The emf generated around it is $\mathcal{E} = -BL\,dx/dt = -BLv$, where v is the speed of the loop. The emf is proportional to the speed.

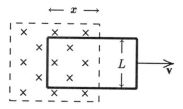

☐ If part of a loop moves in such a way that the area changes, an emf is generated even if the field is uniform. The flux through the loop is a function of time because the area is a function of time. If the field is uniform and perpendicular to the area, the emf is given by $\mathcal{E} = -B\,dA/dt$. It is generated only along the moving part, not around the whole loop.

☐ Part or all of the loop being considered may be imaginary. An emf is generated along an isolated rod moving in a magnetic field, for example. To calculate it, you may complete the loop with imaginary lines. If the rod is conducting, electrons move in the direction opposite to the emf when the emf first appears. They collect at one end of the rod, leaving the other end positively charged, and as a result, an electric field exists along the rod. Very quickly the field prevents further build-up of charge and the current becomes zero. Do not confuse the electric field created by charge at the ends of a moving rod with an electric field associated with a time-varying magnetic field.

☐ An emf is also generated around a loop that is changing orientation so the angle between its normal and the magnetic field is changing with time. Suppose a loop with area A rotates in a uniform magnetic field, perpendicular to the axis of rotation. The flux through the loop is given by $\Phi_B = BA \cos\theta$, where θ is the angle between the normal to the plane of the loop and the magnetic field. If θ is changing according to $\theta = \omega t$, then $\mathcal{E} = -d\Phi_B/dt = \omega AB \sin(\omega t)$.

☐ When a current is induced, energy is transferred, via the emf, to the moving charges. Recall that the rate with which an emf $\mathcal{E}$ does work on a current i is given by $P = \mathcal{E}i$. When the emf is associated with a changing magnetic field, the energy comes from the agent changing the field; when the emf is motional, the energy comes from the agent moving the loop or from the kinetic energy of the loop. In either case, the energy is dissipated in the resistance of the loop.

☐ Consider the example above in which a rectangular loop moves with constant speed out of a region with a uniform magnetic field and recall that the magnitude of the emf generated is given by $\mathcal{E} = BLv$. If the resistance of the loop is R, then the current is given by $i = \mathcal{E}/R = BLv/R$, so the emf transfers energy at the rate $P = B^2L^2v^2/R$.

☐ This is precisely the rate with which the external agent does work on the loop to keep it moving at a constant speed. The external field **B** exerts a net force of magnitude $F = iBL = B^2L^2v/R$ on the loop. The direction of the current is clockwise in the diagram and the magnetic field is into the page so the magnetic force on the moving loop is directed to the left. If the loop is to move with constant velocity, an external agent must apply a force of equal magnitude but in the opposite direction, to the right. Since the loop is moving to the right with speed v, the rate with which the agent does work is $P = Fv = B^2L^2v^2/R$.

☐ If the agent stops pulling, the magnetic field exerts a force that slows the moving loop and eventually stops it. The kinetic energy of the loop is then dissipated in its resistance. This is the basis of magnetic braking.

31–6 Induced Electric Fields

☐ An electric field is always associated with a changing magnetic field and this electric field is responsible for the induced emf. The relationship between the electric field **E** and the emf around a closed loop is given by the integral

$$\mathcal{E} = \oint \mathbf{E} \cdot \mathbf{ds},$$

where **ds** is an infinitesimal displacement vector. The integral is zero for a conservative field, such as the electrostatic field produced by charges at rest. The electric field induced by a changing magnetic field, however, is non-conservative and the integral is not zero. For a changing magnetic field, Faraday's law becomes

$$\oint \mathbf{E} \cdot \mathbf{ds} = -\frac{d\Phi}{dt}.$$

☐ Suppose a cylindrical region of space contains a uniform magnetic field, directed along the axis of the cylinder, and that the field is zero outside the region. If the magnetic field changes with time, the lines of the electric field it induces are circles, concentric with the cylinder.

☐ First, consider a point inside the cylinder, a distance r from the center. The magnetic flux through a circle of radius r is $\pi r^2 B$. The circle coincides with an electric field line and the electric field has uniform magnitude around the circle, so the emf is $\mathcal{E} = 2\pi r E$. Faraday's law gives $2\pi r E = -\pi r^2 \, dB/dt$, so $E = -\frac{1}{2} r \, dB/dt$.

☐ Now consider a point outside the cylinder. The electric field lines are again circles, concentric with the cylinder. Since the region of the magnetic field extends only a distance R from the cylinder center, the magnetic flux through a circle of radius r is $\pi R^2 B$, where R is the radius of the cylinder. Since the emf is again $\mathcal{E} = 2\pi r E$, Faraday's law gives $2\pi r E = -\pi R^2 \, dB/dt$ and $E = -(R^2/2r) \, dB/dt$.

31–7 Inductors and Inductance

☐ Current in a circuit produces a magnetic field and magnetic flux through the circuit. If the circuit consists of N turns and the flux is the same through all of them, then its **inductance** L is defined by

$$L = \frac{N\Phi_B}{i},$$

where i is the current that produces flux Φ_B through each turn. Since Φ_B is proportional to i, L does not depend on the current or flux. It does depend on the geometry of the circuit.

☐ The SI unit of inductance is called the henry (abbreviated H). The quantity $N\Phi$ is called the **flux linkage**.

☐ For an ideal solenoid of length ℓ and cross-sectional area A, with n turns per unit length and carrying current i, the magnetic field in the interior is given by $B = \mu_0 in$, the flux through each turn is given by $\Phi_B = BA = \mu_0 inA$, the flux linkage is $N\Phi_B = \mu_0 in^2 A\ell$, and the inductance is $L = n\ell\Phi_B/i = \mu_0 n^2 A\ell$.

☐ Consider a toroid with N turns, a square cross section, inner radius a, and outer radius b. If the current in the toroid is i, the magnetic field a distance r from the center, within the toroid, is $B = \mu_0 iN/2\pi r$ and the flux through a cross section is

$$\Phi_B = \int_a^b B \, dA = \int_a^b \frac{\mu_0 iN}{2\pi r} h \, dr = \frac{\mu_0 iNh}{2\pi} \ln(b/a).$$

The inductance is $L = N\Phi_B/i = (\mu_0 N^2 h/2\pi)\ln(b/a)$.

31–8 Self-Induction

☐ When the current in a circuit changes, the flux changes and an emf is induced in the circuit. If the circuit has inductance L, the induced emf is

$$\mathcal{E}_L = -L\,\frac{di}{dt}\,.$$

☐ Every circuit has an inductance, usually small, but there are electrical devices, called **inductors**, that are used expressly to add inductance to a circuit. They usually consist of a coil of wire, like a solenoid. The symbol for an inductor is ⟽⟽⟽⟽⟽ .

☐ The diagram below shows an inductor carrying current i, directed from a to b. The rest of the circuit is not shown. If the current is increasing, then the emf induced in the inductor is from b toward a (a is the positive terminal); if it is decreasing, then the emf is from a toward b (b is the positive terminal).

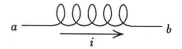

In either case, the potential V_b at point b is given by $V_b = V_a - L\,di/dt$, where V_a is the potential at point a and L is the inductance.

31–9 RL Circuits

☐ Consider a circuit consisting of an inductor L, a resistor R, and an emf $\mathcal{E}$. Take the current to be positive in the direction of the emf. The loop rule then gives $\mathcal{E} - iR - L\,di/dt = 0$. Assume the source of emf $\mathcal{E}$ is connected to

the circuit at time t = 0, at which time the current is 0. The
solution to the loop equation is then

$$i(t) = \frac{\mathcal{E}}{R} \left(1 - e^{-t/\tau_L}\right) ,$$

where $\tau_L = L/R$ is the **inductive time constant**. This expression predicts $i = 0$ at $t = 0$, but di/dt is not zero then.
It also predicts that long after the emf is connected, the current is $i = \mathcal{E}/R$, as if the inductor were not present. The
rate at which the current increases to its final value is controlled by the inductive time constant. The larger the time
constant, the slower the rate.

☐ The potential difference across the inductor is given by

$$V_L(t) = L \frac{di}{dt} = \mathcal{E} e^{-t/\tau_L}$$

and the potential difference across the resistor is given by

$$V_R(t) = iR = \mathcal{E} \left(1 - e^{-t/\tau_L}\right) .$$

These functions are graphed in Fig. 31–21.

☐ The current is the least but its rate of change is the greatest
at $t = 0$. Then, the potential difference across the resistor
is zero and the potential difference across the inductor is $\mathcal{E}$.
A long time after the emf is connected, the current is $\mathcal{E}/R$
and its rate of change is zero. Then, the potential difference
across the resistor is $\mathcal{E}$ and the potential difference across
the inductor is zero.

☐ Suppose that at time $t = 0$, the circuit has a steady current
i_0 and the emf is replaced by a wire. The loop equation is
$L \, di/dt + iR = 0$ and its solution is

$$i(t) = i_0 e^{-t/\tau_L} .$$

The current decreases exponentially to zero. The rate of
decrease is controlled by the inductive time constant.

31–10   Energy Stored in a Magnetic Field

☐   Energy must be supplied to build up the magnetic field in an inductor, perhaps by a source of emf. The energy may be considered to be stored in the magnetic field and can be retrieved when the current and field decrease. If current i is in an inductor with inductance L, the energy stored is given by $U_B = \frac{1}{2}Li^2$.

☐   Multiply the loop equation for a series LR circuit by the current to obtain $i\mathcal{E} = i^2R + Li\,di/dt$. The quantity on the left is the rate with which the emf device is supplying energy to the circuit. The first term on the right is the rate with which energy is dissipated in the resistor. The second term on the right is the rate with which energy is being stored in the magnetic field of the inductor.

31–11   Energy Density of a Magnetic Field

☐   The inductance of a solenoid with cross-sectional area A, length ℓ, and n turns per unit length is $L = \mu_0 n^2 A\ell$ and the energy stored is $U_B = \frac{1}{2}Li^2 = \frac{1}{2}\mu_0 n^2 A\ell i^2$, where i is the current. Since the magnetic field in the solenoid is $B = \mu_0 ni$, this can be written $U_B = B^2 A\ell/2\mu_0$. $A\ell$ is the volume of the solenoid, so the energy density (energy per unit volume) is

$$u_B = \frac{B^2}{2\mu_0}.$$

This expression gives the energy density stored in *any* magnetic field, not just the field of a solenoid. The total energy stored in a field is the volume integral of the energy density.

31–12   Mutual Induction

☐   The current in one circuit influences the current in another, although the two are not physically connected. The current in the first circuit produces a magnetic field at all points in space and thus is responsible for a magnetic flux through the second circuit. When the current in the first circuit changes, the flux through the second circuit changes and an emf is induced in that circuit.

☐ If the second circuit has N_2 turns and Φ_{21} is the flux produced through it by the current i_1 in the first circuit, then

$$M_{21} = \frac{N_2 \Phi_{21}}{i_1}$$

is the **mutual inductance** of circuit 2 with respect to circuit 1. Similarly, the mutual inductance of circuit 1 with respect to circuit 2 is $M_{12} = N_1 \Phi_{12}/i_2$, where N_1 is the number of turns in circuit 1, i_2 is the current in circuit 2, and Φ_{12} is the flux through circuit 1 produced by the current in circuit 2. The two mutual inductances M_{12} and M_{21} are always equal and the subscripts are not needed.

☐ When the current i_1 in circuit 1 changes at the rate di_1/dt, the emf induced in circuit 2 is $\mathcal{E}_2 = -M\,di_1/dt$ and when the current i_2 in circuit 2 changes at the rate di_2/dt, the emf induced in circuit 1 is $\mathcal{E}_1 = -M\,di_2/dt$.

NOTES:

Chapter 32
MAGNETISM OF MATTER; MAXWELL'S EQUATIONS

No magnetic charge, similar to electric charge, has been observed. This statement is codified in Gauss' law for magnetism, one of the fundamental laws of electromagnetic theory. Understand both the mathematical statement and the important ramifications of the law. Magnetic dipoles, not monopoles, are the fundamental magnetic entities responsible for the magnetic properties of matter. You should understand that the motions of electrons in matter produce atomic dipole moments and that the dipole moment of an atom is intimately related to its angular momentum. You should be able to describe the magnetic properties of paramagnetic, diamagnetic, and ferromagnetic materials and you should understand how the behavior of atomic dipoles leads to these properties. Maxwell's equations, the four equations that describe the electric and magnetic fields produced by any source, are presented together in this chapter so you can see how they complement each other. New topics deal with magnetic induction and the displacement current.

Important Concepts

- [ ] Gauss' law for magnetism
- [ ] relationship between
  magnetic dipole moment
  and angular momentum
- [ ] magnetic monopole
- [ ] spin magnetic dipole moment
- [ ] orbital magnetic dipole moment
- [ ] Bohr magneton
- [ ] magnetization
- [ ] diamagnetism
- [ ] paramagnetism
- [ ] Curie's law
- [ ] ferromagnetism
- [ ] Curie temperature
- [ ] hysteresis
- [ ] ferromagnetic domain
- [ ] induced magnetic field
- [ ] Maxwell's law of induction
- [ ] displacement current
- [ ] Maxwell's equations

32–1 Magnets

☐ The magnetic dipole is the simplest known magnetic structure. Magnetism arises from the dipole moments of electrons in materials, associated with their spins and orbital motions.

☐ Magnetic field lines exit a bar magnet from the end called the north pole and enter the end called the south pole. The lines continue through the interior of the magnet to form closed loops.

☐ The magnetic field in the exterior of a permanent bar magnet can be closely approximated by the field that would be produced by a positive monopole (a north pole) at one end and a negative monopole (a south pole) at the other but the field does not actually arise from single monopoles but rather from magnetic dipoles associated with electron motion. As proof that the field is not due to monopoles, the magnet can be cut in half with the result that each half produces a field that is closely approximated as a dipole field. This process can be continued to the atomic level with the same result.

32–2 Gauss' Law for Magnetic Fields

☐ If the normal component of the magnetic field is integrated over a *closed* surface (one that completely surrounds a volume), the result is zero no matter what closed surface is chosen. That is,

$$\oint \mathbf{B} \cdot d\mathbf{A} = 0$$

for every closed surface. The direction of the infinitesimal element of area $d\mathbf{A}$ is normal to the surface. This is Gauss' law for magnetism.

☐ The law does not necessarily mean that the magnetic field is zero at any point on the surface, only that the total magnetic flux through any closed surface is zero. Usually the field is essentially outward over some portions and essentially inward over others. Every magnetic field line that enters any region also leaves that region: no field lines start or stop anywhere; they are closed curves.

☐ The law means that **magnetic monopoles** do not exist, or at any rate are so rare that their influence has not been detected. If monopoles existed and if a closed surface surrounded one of them, the right side of the equation would not be zero. Magnetic field lines would start and stop at monopoles so a net magnetic flux would pass through the surface. On the other hand, magnetic dipoles do exist: a charge circulating around a small loop is an example. They do not violate Gauss' law for magnetism and are, in fact, the fundamental sources of magnetism in matter.

32–3   The Magnetism of Earth

☐ The magnetic field of Earth can be approximated by the field of a magnetic dipole with moment $\mu = 8.0 \times 10^{22}$ J/T, located at Earth's center. The dipole moment is not along the axis of rotation but is tilted slightly from that direction. It points from north to south: magnetic field lines leave Earth in the southern hemisphere and enter it in the northern hemisphere. Thus, the north geomagnetic pole is actually the south pole of Earth's magnetic dipole.

☐ The direction of Earth's magnetic field at any point is often specified in terms of its **declination** and **inclination**. The declination is the angle between the true geographic north and the horizontal component of the magnetic field. The inclination is the angle between the field and a horizontal plane. If the field has horizontal component B_h and vertical component B_v, then the inclination ϕ_i is given by $\tan \phi_i = B_v/B_h$.

32–4   Magnetism and Electrons

☐ Every electron has an intrinsic angular momentum, often called its spin angular momentum or simply its spin. Only one component, usually taken to be the z component, can be measured. It is

$$S_z = \pm \frac{h}{4\pi} = \pm 5.2729 \times 10^{-35}\, \text{J} \cdot \text{s}\,,$$

where h is the Planck constant. A magnetic dipole moment is associated with spin and its z component is

$$\mu_{S,z} = -\frac{e}{m} S_z = \mp \frac{eh}{4\pi m} = \mp 9.27 \times 10^{-24}\, \text{J/T}\,,$$

where m is the electron mass. Since an electron is negatively charged, its dipole moment and spin angular momentum are in opposite directions.

☐ Particle and atomic magnetic moments are often measured in units of the **Bohr magneton** μ_B: $\mu_B = eh/4\pi m$. This is the magnitude of the electron spin dipole moment.

☐ A magnetic moment is also associated with the orbital motion of an electron. The z component of the angular momentum is

$$L_{\text{orb},z} = m_\ell \frac{h}{2\pi}\,,$$

where $m_\ell = \pm 1, \pm 2, \ldots, \pm(\text{limit})$ and "limit" is the largest magnitude of m_ℓ. The z component of the dipole moment is

$$\mu_{\text{orb},z} = -m_\ell \frac{eh}{4\pi} = -m_\ell \mu_B\,.$$

☐ The magnetic energy of an electron in an external magnetic field, in the z direction, is given by $U = -\mu_z B_{\text{ext}}$.

32–5 Magnetic Materials

☐ Three types of magnetic materials (diamagnetic, paramagnetic, and ferromagnetic) are discussed in the following sections. All atoms are diamagnetic. A dipole moment is induced by an external field. Some atoms are paramagnetic. They have permanent dipole moments but these are randomly oriented in the absence of an external field. When an external field is turned on they tend to align with the field. Ferromagnetic atoms also have permanent dipole moments, but they align with each other even in the absence of an external field.

☐ Paramagnetism and ferromagnetism, when they occur, are much stronger than diamagnetism.

32–6 Diamagnetism

☐ In the absence of an applied field, the atoms of a diamagnetic substance have no magnetic dipole moments, but moments are induced when a field is applied. As an external field is turned on, the orbits of the electrons change so the electrons produce an opposing field, in accordance with Faraday's law. The direction of the magnetization (and the induced dipole moments, on average) is opposite that of the local magnetic field, so the total magnetic field in a diamagnetic material is less in magnitude than the applied field. For most diamagnetic materials, the dipole field is extremely weak.

☐ The effect occurs for all materials, but if the atoms have permanent dipole moments, the effect of their alignment with the field dominates and the material is paramagnetic rather than diamagnetic.

☐ If the external field is not uniform, a diamagnetic material is repelled from a region of greater field toward a region of lesser field.

32–7   Paramagnetism

☐   The **magnetization** of a uniformly magnetized object is its magnetic moment per unit volume. The SI unit of magnetization is A/m. If the substance is not uniformly magnetized, the magnetization at a point in the object is the limiting value of the dipole moment per unit volume as the volume shrinks to the point.

☐   Atoms of paramagnetic materials have permanent dipole moments: the dipole moments of the electrons in one of these atoms do not sum to zero. When no external magnetic field is applied, however, the magnetization of the material is zero because the atomic dipole moments are randomly oriented. An external field aligns the atomic moments, and the substance becomes magnetized. When the applied field is removed, the magnetization is quickly reduced to zero by atomic oscillations, which randomize the moments.

☐   When an external magnetic field is applied, the direction of the field produced by dipoles of the material is in the same direction as the applied field, so the total field in the material is greater in magnitude than the applied field.

☐   According to **Curie's law**, the magnetization at any point in a paramagnetic substance is directly proportional to the magnetic field B at that point and inversely proportional to the absolute temperature T:

$$M = C \left(\frac{B}{T} \right) ,$$

where the constant of proportionality C is called the **Curie constant** for the material. The relationship is valid only for small values of B/T. The magnetization increases with the applied field because the stronger the field the more atomic dipoles become aligned. The decrease with temperature reflects the randomizing effect of thermal oscillations.

☐ For large magnetic fields, the magnetization is no longer proportional to the field and, in fact, is less in magnitude than a proportional relationship predicts. For sufficiently high fields, the magnetization is independent of the field. The magnetization is then said to be *saturated*. This occurs when all the dipoles are aligned. If the sample contains N dipoles per unit volume, each with dipole moment μ, then the saturation value of the magnetization is $M_{max} = N\mu$. See Fig. 32–10 for a graph of the magnetization as a function of the magnetic field.

☐ The potential energy of a dipole μ in a magnetic field **B** is given by $U = -\mu \cdot \mathbf{B}$. The energy changes if the orientation of the dipole changes. Suppose a dipole originally aligned with a field is turned end for end. Its initial potential energy is $U_i = -\mu B$, its final potential energy is $U_f = +\mu B$, and the energy required to turn the dipole is $U_f - U_i = 2\mu B$. To see if the thermal energy of the material is adequate to inhibit magnetization, the change in energy is compared with the mean kinetic energy of the atoms of the material. For an ideal gas of atoms, this is $\frac{3}{2}kT$ at absolute temperature T. At room temperature, the thermal energy is typically several hundred times the magnetic energy and is easily sufficient to randomize the dipole orientations.

32–8 Ferromagnetism

☐ Atoms of ferromagnetic materials have permanent dipole moments but unlike the atomic moments of paramagnetic materials, they *spontaneously* align with each other. Iron is the best known ferromagnetic material.

☐ The spontaneous alignment of dipoles in a ferromagnet is *not* due to the magnetic torque exerted by one magnetic dipole on another. These torques are not sufficiently strong to overcome thermal agitation that tends to randomize the dipole orientations. The source of dipole alignment is in the quantum mechanics of electrons in solids.

☐ For temperatures above its **Curie temperature**, a ferromagnetic substance is paramagnetic. For iron, this temperature is 1043 K.

☐ Ferromagnetic materials exhibit **hysteresis**. This may be demonstrated by plotting the magnetic field B_M, due to the atomic dipoles, as a function of the applied field B_0. See Fig. 32–14. If the substance starts in the unmagnetized state and the external field is increased from zero, B_M is nearly linear in B_0 at first but then at higher applied fields, the slope becomes less. Eventually the magnetization becomes saturated and B_M is constant. When the applied field is reduced, B_M does not follow the same curve downward and, in fact, when $B_0 = 0$, B_M is not zero. A residual magnetism remains: the material is magnetized even though there is no external field.

☐ **Ferromagnetic domains** are responsible for hysteresis. All the dipole moments in a domain are aligned; dipoles in neighboring domains are in different directions. The field B_M is the vector sum of the fields due to all domains. If a ferromagnet is unmagnetized, the vector sum of all the dipole moments is zero. When an external field is applied, domains with dipole moments parallel to the field tend to grow in size while domains with dipole moments in other directions tend to shrink. This amounts to changes in the orientations of dipoles near domain boundaries. In some materials, slight re-orientation of dipoles within a domain may also take place.

☐ Hysteresis comes about because the growth and shrinkage of domains is not reversible. When an external field is applied to an unmagnetized sample and then turned off, the domains do not spontaneously revert to their original sizes.

☐ To measure the magnetic field of the atomic dipoles, the core of a toroid is filled with the ferromagnetic material. If the toroid is narrow and has a large radius, the field of the current i_p in it can be approximated by the field of a solenoid: $B_0 = \mu_0 n i$, where n is the number of turns per unit length. The total field is measured by wrapping a secondary coil of wire around the toroid, measuring the current in it when i is changed, and using Faraday's law to find the total field B. Finally, $B = B_0 + B_M$ is used to compute B_M.

32–9 Induced Magnetic Fields

☐ A changing electric field produces a magnetic field. The relationship is

$$\oint \mathbf{B} \cdot d\mathbf{s} = \mu_0 \epsilon_0 \frac{d\Phi_E}{dt}.$$

The left side is a path integral around a closed path. Φ_E is the electric flux through the area bounded by the path. In this form, the equation is called the **Maxwell law of induction**.

☐ A sign convention is associated with the Maxwell induction law. First, pick the direction of $d\mathbf{A}$ to be used to compute the electric flux. It is usually chosen to make the flux positive. Then, point the thumb of your right hand in the direction of $d\mathbf{A}$; your fingers then curl around the boundary in the direction of $d\mathbf{s}$.

☐ Suppose a uniform but time-varying electric field exists in a cylindrical region of radius R and is parallel to the axis of the cylinder. The magnetic field lines are circles that are concentric with the cylinder and the magnetic field is uniform on any one of them. For the field line with radius r, $\oint \mathbf{B} \cdot d\mathbf{s} = 2\pi r B$. If the circle is inside the cylinder ($r < R$), then $\Phi_E = \pi r^2 E$ and the Maxwell induction law gives $2\pi r B = \pi r^2 \mu_0 \epsilon_0\, dE/dt$, so $B = \frac{1}{2}\mu_0 \epsilon_0 r\, dE/dt$. If the circle is outside the cylinder ($r > R$), then $\Phi_E = \pi R^2 E$ and the Maxwell induction law gives $2\pi r B = \pi R^2 \mu_0 \epsilon_0 dE/dt$, so $B = (\mu_0 \epsilon_0 R^2 / r)\, dE/dt$. The magnetic field increases linearly with distance away from the center of the cylinder until the boundary is reached, then it decreases like $1/r$.

☐ Currents also produce magnetic fields, as you learned when you studied Ampere's law. This law and the Maxwell induction law can be combined to yield

$$\oint \mathbf{B} \cdot d\mathbf{s} = \mu_0 i + \mu_0 \epsilon_0 \frac{d\Phi_E}{dt}.$$

32–10 Displacement Current

☐ According to the Ampere-Maxwell law, a changing electric field in a region produces a magnetic field around the boundary of the region, just as if a current passed through the region. In fact, the quantity

$$i_d = \epsilon_0 \frac{d\Phi_E}{dt}$$

is called a **displacement current**. A displacement current is emphatically NOT a true current, which consists of moving charges.

☐ For a cylindrical region of radius R containing a uniform changing electric field along its axis, the displacement current through a circular loop of radius r is $i_d = \epsilon_0 \pi r^2 \, dE/dt$ if $r < R$ and is $i_d = \epsilon_0 \pi R^2 \, dE/dt$ is $r > R$. The direction of the displacement current is the same as the direction of the field if E is increasing and opposite the direction of the field if E is decreasing.

☐ The equations for the magnetic field associated with the cylindrical region considered above are just like the equations for the magnetic field of a long straight wire, developed in Chapter 30, but the true current i is replaced by the displacement current i_d. That is, the field inside the cylinder is $B = \mu_0 i_d (r^2/R^2)/2\pi r$ and the field outside the cylinder is $B = \mu_0 i_d / 2\pi r$, where i_d is the total displacement current in the cylinder.

☐ A charging (or discharging) capacitor provides an example of a displacement current. The charge on the plates produces an electric field in the interior and, since the charge is changing, the field is also. Consider a parallel-plate capacitor with plate area A and let $q(t)$ be the charge on the positive plate. Then, the magnitude of the electric field between the plates is $E(t) = q(t)/\epsilon_0 A$. Consider a cross section between the plates with the same area as a plate. The electric flux through the cross section is $\Phi_E = EA$ and the total displacement current through it is $i_d = \epsilon_0 A \, \mathrm{d}E/\mathrm{d}t = \mathrm{d}q/\mathrm{d}t$. That is, the total displacement current in the interior is the same as the current in the capacitor wires. The sum of the true and displacement currents is continuous. Although the true current stops at the plates, the displacement current continues into the interior.

32–11 Maxwell's Equations

☐ Here is list of Maxwell's equations, including the displacement current term in the Ampere-Maxwell law:

Gauss' law for electricity: $\oint \mathbf{E} \cdot \mathrm{d}\mathbf{A} = \dfrac{q}{\epsilon_0}$.

Gauss' law for magnetism: $\oint \mathbf{B} \cdot \mathrm{d}\mathbf{A} = 0$.

Faraday's law of induction: $\oint \mathbf{E} \cdot \mathrm{d}\mathbf{s} = -\dfrac{\mathrm{d}\Phi_B}{\mathrm{d}t}$.

Ampere-Maxwell law: $\oint \mathbf{B} \cdot \mathrm{d}\mathbf{s} = \mu_0 i + \mu_0 \epsilon_0 \dfrac{\mathrm{d}\Phi_E}{\mathrm{d}t}$.

☐ The left sides of Gauss' law for electricity and Gauss' law for magnetism contain integrals over closed surfaces. The symbol q on the right side of the law for electricity represents the net charge enclosed by the surface. The zero on the right side of the law for magnetism indicates that magnetic monopoles do not exist. The left sides of Faraday's law and the Ampere-Maxwell law contain integrals around closed paths. On the right side of Faraday's law, Φ_B is the magnetic flux through the surface bounded by the path. On the right side of the Ampere-Maxwell law, i is the net current through the surface bounded by the path and Φ_E is the total electric flux through that surface.

☐ We used Gauss' law for electricity to find the electric field of various charge distributions; we used Faraday's law to find the electric field produced by a changing magnetic field; and we used the Ampere-Maxwell law to find the magnetic field produced by currents and by changing electric fields.

NOTES:

Chapter 33
ELECTROMAGNETIC OSCILLATIONS AND ALTERNATING CURRENT

The combination of a charged capacitor and an inductor connected in series produces a sinusoidally varying current. Learn what factors determine the frequency of oscillation. When a resistor is added, the amplitude decays exponentially with time. Energy is an important theme of this chapter. Pay careful attention to the form it takes, sometimes as electrical energy in the capacitor and sometimes as magnetic energy stored in the inductor. In the first, circuit energy is conserved but when a resistor is added, it is dissipated as thermal energy. You then study a series circuit consisting of a resistor, an inductor, and a capacitor, driven by a sinusoidal emf. The current is an alternating current (ac): it is sinusoidal and periodically reverses direction. Pay close attention to the relationship between the potential difference across each circuit element and the current in the element. Learn how to compute the current amplitude and phase in terms of the generator emf. Also learn about electrical power in an ac circuit.

Important Concepts

- [ ] *LC* oscillations
- [ ] charge amplitude
- [ ] current amplitude
- [ ] electrical-mechanical analogy
- [ ] damped electromagnetic oscillations
- [ ] alternating current

- [ ] capacitive reactance
- [ ] inductive reactance
- [ ] impedance
- [ ] resonance
- [ ] power factor
- [ ] transformer

33–1   New Physics — Old Mathematics

☐ The mathematics of oscillations was discussed in Chapter 16 in connection with a block-spring system. The *currents* in the circuits of this chapter oscillate; the physics is different but the mathematics is the same.

33–2   *LC* Oscillations, Qualitatively

☐ If the circuit consists only of an inductor with inductance L and a capacitor with capacitance C, the charge on either plate of the capacitor oscillates sinusoidally, being positive for half a cycle and negative for the other half. The current also oscillates, being in one direction for half a cycle and in the other direction for the other half. The energy also oscillates between the capacitor and the inductor.

☐ Study Fig. 33–1 carefully. Suppose that initially the charge on the capacitor is Q, with the upper plate positive, and that the current is zero. The energy in the circuit is $Q^2/2C$ and it is all in the capacitor. Immediately, there is a current away from the upper plate, driven by the potential difference across the capacitor. As time goes on, the current increases and the charge on the capacitor decreases; the magnetic field in the inductor increases; the energy in the capacitor decreases; and the energy in the inductor increases. After one-fourth of a cycle, the charge on the capacitor is zero and the current is a maximum. Then, all the energy resides in the inductor.

☐ The current continues, so the lower plate of the capacitor becomes positively charged. The charge there causes the current to decrease. Energy in the capacitor increases and energy in the inductor decreases. At half a cycle, the capacitor charge is again Q, but with the lower plate positive, and the current is zero. All the energy now resides in the capacitor.

☐ The process is now reversed, with the current away from the lower plate of the capacitor. At the end of a cycle, the capacitor again has charge Q, with the upper plate positive, and the process repeats.

☐ Since energy is conserved, $\frac{1}{2}LI^2 = Q^2/2C$, where I is the maximum current. As a result, the maximum current and the maximum charge on the capacitor are related by $I = Q/\sqrt{LC}$.

33–3 The Electrical – Mechanical Analogy

☐ The electromagnetic oscillations of an LC circuit are quite similar to the mechanical oscillations of a block-spring system. The kinetic energy of the block is $\frac{1}{2}mv^2$ and the magnetic energy of the inductor is $\frac{1}{2}Li^2$; the potential energy of the spring is $\frac{1}{2}kx^2$ and the electric energy of the capacitor is $q^2/2C$. Here k is the spring constant, m is the mass of the block, x is its coordinate, and v is its speed. The energy equations for the block-spring system can be turned into the energy equations for the LC circuit if x is replaced by q, v is replaced by i, k is replaced by $1/C$, and m is replaced by L.

☐ Since the angular frequency of the block is $\omega = \sqrt{k/m}$, the replacements suggest that the angular frequency of the LC circuit should be $\omega = 1/\sqrt{LC}$. It is.

33–4 LC Oscillations, Quantitatively

☐ The first step is to develop and solve the loop equation for the circuit. Let q represent the charge on the upper plate of the capacitor and take the current to be positive if it is into that plate. Either Kirchhoff's loop rule or differentiation with respect to time of the energy conservation equation $U = \frac{1}{2}Li^2 + q^2/2C = \text{constant}$ yields

$$L\frac{d^2q}{dt^2} + \frac{1}{C}q = 0,$$

after di/dt is replaced with d^2q/dt^2.

□ The solution to this equation is

$$q(t) = Q \cos(\omega t + \phi),$$

where $\omega = 1/\sqrt{LC}$. Q is the **charge amplitude**, the maximum charge on the capacitor, and ϕ is a phase constant that is determined by the initial conditions. The current is given by

$$i(t) = \frac{dq}{dt} = -\omega Q \sin(\omega t + \phi).$$

Both q and i oscillate with an angular frequency that is determined by the values of L and C. The **current amplitude** I is ωQ.

□ The current is zero when the charge on the capacitor is a maximum, either $-Q$ or $+Q$. The charge on the capacitor is zero when the current is a maximum, either $-\omega Q$ or $+\omega Q$.

□ The electrical energy in the capacitor is

$$U_E = \frac{q^2}{2C} = \frac{Q^2}{2C} \cos^2(\omega t + \phi).$$

It is a maximum when the charge on the capacitor is a maximum; it is zero when the charge is zero. The magnetic energy in the inductor is

$$U_B = \tfrac{1}{2}Li^2 = \tfrac{1}{2}L\omega^2 Q^2 \sin^2(\omega t + \phi).$$

It is zero when the current is zero and the charge on the capacitor is a maximum. The electric energy is then a maximum. The magnetic energy is a maximum when the current is a maximum and the charge on the capacitor is zero. The electric energy is then zero.

□ Because $\omega^2 = 1/LC$, the expression for the magnetic energy can be written $U_B = (Q^2/2C) \sin^2(\omega t + \phi)$. The maximum magnetic energy has the same value as the maximum electric energy. Energy is conserved: $U_E + U_B = Q^2/2C$, a constant.

33–5 Damped Oscillations in an RLC Circuit

☐ If a resistor with resistance R is added in series to the capacitor and inductor, the charge and current amplitudes decay exponentially with time. The loop equation becomes

$$L \frac{d^2q}{dt^2} + R \frac{dq}{dt} + \frac{1}{C} q = 0$$

and its solution is

$$q(t) = Q \, e^{-Rt/2L} \cos(\omega' t + \phi),$$

where the angular frequency of oscillation is now

$$\omega' = \sqrt{\omega^2 - (R/2L)^2}.$$

Each time the charge on the capacitor reaches a maximum, it is less than the previous time. The sum of the electric and magnetic energies is not constant but decreases exponentially:

$$U = \frac{Q^2}{2C} e^{-Rt/L}.$$

Energy is dissipated by the resistor.

33–6 Alternating Current

☐ Alternating, rather than direct, current is used for power transmission because the current and potential difference can easily be increased or deceased by means of a transformer, which relies on changing magnetic flux for its operation. Power is transmitted at high potential difference and low current to reduce dissipation in the resistance of the transmission lines, but then the potential difference is reduced for safety at the user end.

☐ The circuit you consider is diagramed below. The ac generator produces an emf $\mathcal{E} = \mathcal{E}_m \sin(\omega_d t)$, where ω_d is its angular frequency. $\mathcal{E}$ is positive if the upper terminal of the generator is positive and negative if the upper terminal is negative. The arrow shows the direction of positive current.

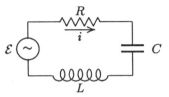

☐ The current is given by $i(t) = I \sin(\omega_d t - \phi)$, where I is the current amplitude and ϕ is a phase constant. Note that the current has the same frequency as the generator emf but because the circuit contains a capacitor and an inductor, it may not be in phase with the generator emf.

33–8 Three Simple Circuits

☐ Quantities with sinusoidal time dependence can be represented by **phasors**. The phasor for the current $i(t) = I \sin(\omega_d t - \phi)$ is a line with length proportional to I that rotates counterclockwise with angular velocity ω_d around the origin. At time $t = 0$, its angular position is $-\phi$. The projection of the phasor on the vertical axis is $I \sin(\omega_d t - \phi)$, the time-dependent current.

☐ Suppose the resistor alone is connected across the generator. Then, the potential difference v_R across the resistor is $v_R = iR = I_R R \sin(\omega_d t)$. v_R and i are in phase; the phasor associated with v_R is on the same line as the phasor associated with i and its length is $V_R = I_R R$.

☐ Suppose a capacitor alone is connected across the generator and take the potential difference across it to be $v_C(t) = V_C \sin(\omega_d t)$. Then, the charge on the capacitor is $q(t) =$

$CV_C \sin(\omega_d t)$, where C is the capacitance. The current into the capacitor is

$$i_C(t) = \frac{dq}{dt} = \omega_d C V_C \cos(\omega_d t) = \omega_d C V_C \sin(\omega_d t + 90°).$$

The trigonometric identity $\sin(\omega_d t + 90°) = \cos(\omega_d t)$ was used. The equation for the current can be written $i_C = I_C \sin(\omega_d t + 90°)$, where $I_C = \omega_d C V_C$. i_C *leads* v_C by 90°. The phasor associated with i_C is 90° ahead of the phasor associated with v_C and its length is $I_C = V_C/X_C$, where $X_C = 1/\omega_d C$. X_C is called the **capacitive reactance** of the capacitor.

☐ Suppose an inductor alone is connected across the generator and take the potential difference across it to be $v_L(t) = V_L \sin(\omega_d t)$. Then, the current through the inductor is

$$i_L = -\int \frac{v_L}{L}\, dt = -\frac{V_L}{\omega_d L} \cos(\omega_d t) = \frac{V_L}{\omega_d L} \sin(\omega_d t - 90°).$$

The trigonometric identity $\sin(\omega_d t - 90°) = -\cos(\omega_d t)$ was used. The equation for the current can be written $i_L = I_L \sin(\omega_d t - 90°)$, where $I_L = V_L/\omega_d L$. i_L *lags* v_L by 90°. The phasor associated with i_L is 90° behind the phasor associated with v_L and its length is $I_L = V_L/X_L$, where $X_L = \omega_d L$. X_L is called the **inductive reactance** of the inductor.

☐ The current-potential difference relationships $V_R = I_R R$ for a resistor, $V_C = I_C X_C$ for a capacitor, and $V_L = I_L X_L$ for an inductor, where $X_C = 1/\omega_d C$ and $X_L = \omega_d L$, are generally valid. In addition, the current in an inductor always lags the potential difference across it by 90° and the current into a capacitor always leads the potential difference across it by 90°. Reactances depend on the generator frequency: as the frequency increases, X_C decreases and X_L increases. X_L, X_C, and R all have SI units of ohms.

33–9 The Series RLC Circuit

☐ Consider the series RLC circuit shown above. According to the loop equation for the circuit $\mathcal{E}(t) = v_R(t) + v_L(t) + v_C(t)$. This means the phasors associated with v_R, v_L, and v_C must add like vectors to produce the phasor associated with $\mathcal{E}$. Since the phasors for v_L and v_C are parallel to each other, the phasor associated with $v_L + v_C$ has length $|V_L - V_C|$. It is 90° ahead of the phasor associated with i if $V_L > V_C$ and 90° behind if $V_C > V_L$. The phasors associated with v_R and $v_L + v_C$ form two sides of a right triangle with a hypotenuse of length $\mathcal{E}_m$, so $\mathcal{E}_m^2 = V_R^2 + (V_L - V_C)^2$. Use $V_R = IR$, $V_L = IX_L$, and $V_C = IX_C$ to write $\mathcal{E} = IZ$, where $Z = \sqrt{R^2 + (X_L - X_C)^2}$ is called the **impedance** of the circuit.

☐ The phasors associated with $\mathcal{E}$ and v_R are separated by ϕ. Use the triangle formed by the phasors associated with $\mathcal{E}$, v_R, and $v_L + v_C$ to show that $\tan \phi = (V_L - V_C)/V_R$. This can be written $\tan \phi = (X_L - X_C)/R$.

☐ The current amplitude I depends on the difference $\omega_d - \omega_0$ between the generator angular frequency ω_d and the natural angular frequency ω_0 ($= 1/\sqrt{LC}$) of the circuit. See Fig. 33–13. As the generator frequency approaches the natural frequency from either side, the current amplitude grows and it reaches its maximum value when $\omega = \omega_0$. The circuit is then said to be at **resonance**. The resonance peak can be made sharper by reducing the resistance in the circuit.

☐ The impedance has its smallest value at resonance. Then, the capacitive and inductive reactances cancel each other and the current is in phase with the generator emf.

33–10 Power in Alternating-Current Circuits

☐ The rate with which energy is supplied by the generator is given by $P_\mathcal{E} = i\mathcal{E}$, the rate at which energy is stored in the capacitor is given by $P_C = iv_C$, the rate at which energy is stored in the inductor is given by $P_L = iv_L$, and the rate at which energy is dissipated in the resistor is given by $P_R = iv_R$. All of these vary sinusoidally with time.

☐ Usually the time variations are of no interest. Consider instead averages over a cycle of oscillation. The average of $\sin^2(\omega_d t - \phi)$ is $1/2$ and the average of $\sin(\omega_d t - \phi)\cos(\omega_d t - \phi)$ is 0. These average values can be used to show that the averages of P_C and P_L are zero and that the average of $P_\mathcal{E}$ is $P_{av} = \frac{1}{2}I\mathcal{E}_m\cos\phi$. If $\mathcal{E}_m = IZ$ is used, this can also be written $P_{av} = \frac{1}{2}(\mathcal{E}_m^2/Z)\cos\phi$ or $P_{av} = \frac{1}{2}I^2 Z\cos\phi$. The quantity $\cos\phi$ is called the **power factor**.

☐ The triangle formed by the phasors associated with $\mathcal{E}$, v_R, and $v_L + v_C$ can be used to show that $\cos\phi = v_R/\mathcal{E} = R/Z$. Thus, the average power supplied by the generator is also given by $P_{av} = \frac{1}{2}I^2 R$.

☐ The rate at which energy is dissipated in the resistor is given by $P_R = i^2 R = I^2 R\sin^2(\omega_d t - \phi)$ and its average value is $P_{av} = \frac{1}{2}I^2 R$, the same as the power supplied by the generator. All energy supplied by the generator is dissipated in the resistor.

☐ For a given emf, the greatest power dissipation occurs when the power factor has the value $\cos\phi = 1$. The impedance must be $Z = R$ and the reactances must be equal: $X_L = X_C$. Since $X_L = \omega_d L$ and $X_C = 1/\omega_d C$, this means $\omega_d^2 = 1/LC$. The circuit is in resonance.

☐ Average power is often expressed in terms of root-mean-square quantities instead of amplitudes. The rms value of a sinusoidal function of time is the amplitude divided by $\sqrt{2}$. For example, the rms value of $\mathcal{E}(t) = \mathcal{E}_m\sin(\omega_d t)$ is $\mathcal{E}_{rms} = \mathcal{E}/\sqrt{2}$. The average power supplied by the generator is $P_{av} = \mathcal{E}_{rms}I_{rms}\cos\phi$.

33–11 Transformers

☐ A **transformer** consists of two coils wrapped around the same iron core. Current is supplied to the primary coil by a generator and current in the secondary coil passes through a load resistance, such as a heater or a motor. Consider an ideal transformer (negligible resistance and hysteresis loss) with N_p turns in the primary coil and N_s turns in the secondary coil. The iron core ensures that the magnetic flux is the same through the two coils, so when the currents change the same emf is induced in the coils. If V_p is the potential difference across the primary coil and V_s is the potential difference across the secondary coil, then $V_p/N_p = V_s/N_s$. If $N_s > N_p$, then $V_s > V_p$ and the transformer is called a *step-up* transformer. If $N_s < N_p$, then $V_s < V_p$ and the transformer is called a *step-down* transformer.

☐ Since there are no losses in an ideal transformer, the rate at which energy enters through the primary coil equals the rate at which energy leaves through the secondary coil and $I_p V_p = I_s V_s$. These are rms values.

☐ If the secondary circuit has resistance R and no reactance, the current in that circuit is $I_s = V_s/R = N_s V_p/N_p R$ and the current in the primary circuit is $I_p = I_s V_s/V_p = N_s^2 V_p/N_p^2 R$, where $V_s = N_s V_p/N_p$ was used.

☐ The current in the primary circuit does not change if the transformer and the secondary circuit are replaced by a resistor with resistance $R_{eq} = V_p/I_p = (N_p/N_s)^2 R$. Transformers are often used to match the impedances of generators (or amplifiers) and their loads. That is, they are used to make R_{eq} the same as the generator resistance. Maximum energy transfer occurs if this condition is met.

NOTES:

Chapter 34
ELECTROMAGNETIC WAVES

Maxwell's equations predict the possibility of electric and magnetic fields that propagate in free space, with the changing magnetic field producing changes in the electric field, via Faraday's law, and the changing electric field producing changes in the magnetic field, via the displacement current term in the Ampere-Maxwell law. Pay close attention to the relationship between the amplitudes of the fields, to the relationship between their phases, and to the relationship between their directions. Learn about the energy and momentum carried by electromagnetic waves. Also learn how to determine the propagation directions of light that is reflected or refracted at the boundary between two materials.

Important Concepts

☐ electromagnetic wave
☐ speed of light
☐ Poynting vector
☐ intensity
☐ radiation pressure
☐ polarization
☐ polarization direction

☐ law of Malus
☐ wave optics
☐ geometrical optics
☐ law of reflection
☐ law of refraction
☐ total internal reflection
☐ Brewster's angle

34–1 Maxwell's Rainbow

☐ Electromagnetic waves exist for all wavelengths and frequencies, from very large to very small. Various ranges of wavelengths have been named: gamma radiation, x-ray radiation, ultraviolet radiation, visible radiation, infrared radiation, and microwave radiation. All these electromagnetic radiations are exactly alike except for their frequencies and wavelengths. They all consist of traveling electric and magnetic fields and they all travel in free space with the same speed, the speed of light $c = 2.9979 \times 10^8$ m/s. The ranges do not have sharp boundaries; they blend into each other.

34–2   The Traveling Electromagnetic Wave, Qualitatively

☐ Classically, electromagnetic waves are produced by accelerating charges. Charges at rest or moving with constant velocity do not radiate. In the quantum mechanical picture, electromagnetic radiation is produced when a charge (perhaps an electron in an atom) changes its state or in reactions of fundamental particles.

☐ Figs. 34–4 and 5 show the electric and magnetic fields of a radiating antenna consisting of straight wires in which the current varies sinusoidally, driven by an LCR circuit. The electric field in the antenna varies sinusoidally and because its current varies sinusoidally, the magnetic field it produces also varies sinusoidally. The fields produced by the antenna travel outward from it with the speed of light.

☐ A traveling electromagnetic wave has the following properties:

1. The electric and magnetic fields are perpendicular to each other.

2. The wave travels in a direction that is perpendicular to both the electric and magnetic fields.

3. The electric and magnetic fields are in phase. Both of them reach their maximum values at the same time and are zero at the same time.

☐ Very far from the source, the wave becomes a plane wave: the electric and magnetic field lines are nearly straight lines. Mathematically, the fields of an electromagnetic wave traveling in the positive x direction are then given by

$$E = E_m \sin(kx - \omega t) \quad \text{and} \quad B = B_m \sin(kx - \omega t),$$

where $k = 2\pi/\lambda$, λ is the wavelength, and ω is the angular frequency. These quantities are related by $\omega/k = c$, where c is the speed of light. The phase constants are the same for the two fields.

☐ The electric field of an electromagnetic wave is the source, via the Ampere-Maxwell law, of the magnetic field and the magnetic field is the source, via Faraday's law, of the electric field. Once started, the fields produce each other and travel together through space. When these laws are applied, you find that the amplitudes of the fields are related by

$$\frac{E_m}{B_m} = c$$

and that the wave speed is determined by the permittivity and permeability constants ϵ_0 and μ_0:

$$c = \frac{1}{\sqrt{\mu_0 \epsilon_0}}.$$

34–3 The Traveling Electromagnetic Wave, Quantitatively

☐ For a wave traveling along the x axis, the electric field is different at a point with coordinate x and a point with coordinate $x + dx$, an infinitesimal distance away, because a changing magnetic flux penetrates the region between these points. The magnetic field is different at x and $x + dx$ because a changing electric flux penetrates the region between these points. Differences in the fields at different points can be computed using the Faraday and Ampere-Maxwell laws.

☐ The following diagram shows a small region of space in which a wave is traveling toward the right. The electric fields at two infinitesimally separated points, x and $x + \Delta x$, are shown, as is the magnetic field in the region between. Apply Faraday's law to this situation. To calculate the magnetic flux, take dA to be out of the page and to calculate the emf, traverse the path shown in the counterclockwise direction. Clearly, $\Phi_B = Bh\Delta x$ and $\oint \mathbf{E} \cdot d\mathbf{s} = h[E(x + \Delta x) - E(x)]$. As Δx becomes small, $[E(x + \Delta x) - E(x)]/\Delta x$ becomes $\partial E/\partial x$ and Faraday's law yields $\partial E/\partial x = -\partial B/\partial t$.

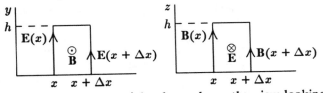

☐ The diagram on the right above shows the view looking up from underneath the previous diagram. The magnetic field at the two points and the electric field between are shown. Apply the Ampere-Maxwell law to this situation. Take dA to be into the page and traverse the path in the clockwise direction. Clearly, $\Phi_E = Eh\Delta x$ and $\oint \mathbf{B} \cdot d\mathbf{s} = h[B(x) - B(x + \Delta x)]$. As Δx becomes small, $[B(x) - B(x + \Delta x)]/\Delta x$ becomes $-\partial B/\partial x$ and the Ampere-Maxwell law yields $\partial B/\partial x = \mu_0\epsilon_0 \, \partial E/\partial t$.

☐ Substitute $E = E_m \sin(kx - \omega t)$ and $B = B_m \sin(kx - \omega t)$ into these two equations and show that $kE_m = \omega B_m$ and that $kB_m = \mu_0\epsilon_0\omega E_m$. Then, use $\omega/k = c$ to show that $E_m = cB_m$ and $c^2 = 1/\mu_0\epsilon_0$.

34–4 Energy Transport and the Poynting Vector

☐ Electromagnetic waves carry energy. The energy per unit volume associated with an electric field $\mathbf{E}$ is given by $u_E = \frac{1}{2}\epsilon_0 E^2$ and the energy per unit volume associated with a magnetic field $\mathbf{B}$ is given by $u_B = B^2/2\mu_0$. You can use $B = E/c$ and $c = 1/\sqrt{\epsilon_0\mu_0}$ to show that for a plane wave $u_E = u_B = EB/2\mu_0 c$ and that the total energy density is $u = EB/\mu_0 c$.

☐ This energy moves with the wave, with speed c. Consider a region of space with infinitesimal width dx and cross-sectional area A, perpendicular to the direction of travel of a plane electromagnetic wave. The volume of the region is $A\,dx$, so the electromagnetic energy in the region is $dU = uA\,dx$. All this energy will pass through the area A in time $dt = dx/c$. Substitute $dx = c\,dt$ and divide by dt. The energy passing through the area per unit time is $dU/dt = ucA$ and the energy per unit area passing through per unit time is $dU/A\,dt = uc = EB/\mu_0$.

☐ The transport of energy is described in terms of the **Poynting vector S**, defined by the vector product

$$\mathbf{S} = \frac{\mathbf{E} \times \mathbf{B}}{\mu_0}.$$

Since **E** and **B** are perpendicular to each other, the magnitude of the Poynting vector is given by $S = EB/\mu_0$. The energy passing through a surface of area A, perpendicular to the direction of travel, per unit time, is given by $dU/dt = SA$ and the energy passing through per unit area, per unit time, is given by $dU/A\,dt = S$.

☐ For most sinusoidal electromagnetic waves of interest, the fields oscillate so rapidly that their instantaneous values cannot be detected or else are not of interest. Energies and energy densities are then characterized by their average over a period of oscillation. For a sinusoidal wave, the average value of E^2 is $\frac{1}{2}E_m^2$, where E_m is the amplitude. Since the rms value of E is $E_{\text{rms}} = E_m/\sqrt{2}$ the average of E^2 is also E_{rms}^2.

☐ The **intensity** of a wave is the magnitude of the Poynting vector averaged over a period of oscillation:

$$I = \frac{E_m B_m}{2\mu_0}.$$

This can be written as $I = E_m^2/2c\mu_0$ and as $I = cB_m^2/2\mu_0$. For sinusoidal waves, the intensity is often written in terms of the rms value E_{rms} of the electric field: $I = E_{\text{rms}}^2/c\mu_0$.

☐ The direction of **S** is the direction in which the wave is traveling. Recall that this direction is intimately connected with the relative signs of the terms kx and ωt in the argument of the trigonometric function that describes the fields. If the wave travels in the negative x direction, you write $E_y(x,t) = E_m \sin(kx + \omega t)$ and $B_z(x,t) = -B_m \sin(kx + \omega t)$. The negative sign appears so that **E** × **B** is in the negative x direction.

34–5 Radiation Pressure

☐ Electromagnetic waves also carry momentum. If ΔU is the energy in any small volume, then $\Delta p = \Delta U/c$ is the magnitude of the momentum in that volume. The momentum is in the direction of **S**, the direction of propagation. If u is the energy per unit volume, then u/c is the momentum per unit volume and if I is the average energy transported through an area per unit area per unit time (the intensity), then I/c is the average momentum transported through the area per unit area per unit time.

☐ The electric field component of an electromagnetic wave exerts a force on a charge and does work on it. If the charge is moving, the magnetic field component also exerts a force. Thus, momentum and energy may be transferred to an object when radiation interacts with it.

☐ When electromagnetic radiation is absorbed by a material object, all its energy and momentum are transferred to the object. Consider a plane wave with time averaged energy density u, incident normally on the plane surface of an object. If the area of the surface is A, then, averaged over a period, the energy transferred in time Δt is given by $IA\,\Delta t$ and the momentum transferred is given by $(IA/c)\,\Delta t$. The momentum transferred per unit time is the force of the radiation on the object and the force per unit area is the **radiation pressure**.

☐ If the wave is reflected without loss back along the original path, the energy transferred is zero and the momentum transferred is $2(IA/c)\,\Delta t$. The radiation pressure is now $2I/c$, twice what it would be if the radiation were absorbed.

34–6   Polarization

☐ The electric field of a **linearly polarized** electromagnetic wave is always parallel to the same direction. The **direction of polarization** is parallel to the electric field and the **plane of polarization** is determined by the directions of propagation and the electric field. For maximum signal, the wires in an electric dipole antenna used to detect polarized waves must be aligned with the polarization direction.

☐ An electromagnetic wave is said to be **unpolarized** if the electric field at any point changes direction randomly and often. Light from an incandescent bulb or from the Sun is unpolarized. The direction of its electric field changes rapidly in a random fashion because the light comes from many different atoms, each of which emit waves in short bursts, with random polarization directions.

☐ Any linearly polarized wave can be treated as the sum of two waves, polarized in any two mutually orthogonal directions that are perpendicular to the direction of propagation. Suppose a wave with an electric field amplitude E_m is traveling out of the page and is polarized with its electric field along a line that makes an angle θ with the x axis. You can consider the electric field to be the vector sum of two fields, one with amplitude $E_m \cos\theta$, polarized along the x axis and one with amplitude $E_m \sin\theta$, polarized along the y axis.

☐ Polarized radiation can be produced by shining unpolarized radiation through a sheet of Polaroid. These sheets contain certain long-chain molecules that are aligned in the manufacturing process. Radiation with its electric field vector parallel to the chains is preferentially absorbed while radiation with its electric field vector perpendicular to the chains is preferentially transmitted. A line perpendicular to the molecules is said to be along the **polarizing direction** of the sheet.

☐ Suppose that linearly polarized radiation with amplitude E_m and intensity I_m is incident on an ideal polarizing sheet and that its electric field is along a line that makes an angle θ with the polarizing direction. Then, the amplitude of the transmitted radiation is given by $E = E_m \cos \theta$ and its intensity is given by $I = I_m \cos^2 \theta$. This is called the **law of Malus**. Note that the transmitted amplitude is the component of the incident amplitude along the polarization direction of the sheet. The transmitted radiation is polarized along the polarizing direction of the sheet.

☐ If unpolarized radiation with intensity I_i is incident on an ideal polarizing sheet, the intensity of the transmitted radiation is given by $I = I_i/2$. The transmitted radiation is polarized along the polarizing direction of the sheet.

34–7   Reflection and Refraction

☐ In **wave optics**, the wave nature of light is used to describe and understand optical phenomena. Wave optics is close to the fundamental principles, Maxwell's equations, and is valid for all optical phenomena outside the quantum realm. When any obstacles to light or any openings through which light passes are large compared to the wavelength of the light, then details of its wave nature are not important. The direction of travel of the light is important. This is the realm of **geometrical optics** (or ray optics).

☐ Geometrical optics uses rays to describe the path of light. A ray is a line in the direction of propagation of a light wave. The propagation direction changes when light is reflected or enters a region where the wave speed is different.

☐ When light in one medium encounters a boundary with another region, in which the wave speed is different, some light is reflected back into the region of incidence and some is transmitted into the second region. The following diagram shows a ray in medium 1 incident on the boundary with medium 2. It makes the angle θ_1 with the normal to the boundary. The reflected ray makes the angle θ_1' with the normal and the transmitted ray makes the angle θ_2 with the normal. The angle θ_1 is called the **angle of incidence**, θ_1' is called the **angle of reflection**, and θ_2 is called the **angle of refraction**. Reflected and refracted rays all lie in the plane of incidence, defined by the incident ray and the normal to the surface. The **law of reflection** is $\theta_1' = \theta_1$ and the **law of refraction** is $n_2 \sin \theta_2 = n_1 \sin \theta_1$, where n_1 is the **index of refraction** for medium 1 and n_2 is the index of refraction for medium 2. The law of refraction is also known as Snell's law.

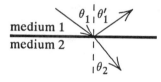

☐ The index of refraction of a medium is the ratio of the speed of light c in vacuum to the speed of light v in the medium: $n = c/v$.

☐ Light incident from a medium with a small index of refraction bends toward the normal as it enters a medium with a higher index of refraction. Light incident from a medium with a large index of refraction bends away from the normal as it enters a medium with a smaller index of refraction.

☐ The index of refraction depends on the wavelength of the light. Light rays of different wavelengths are bent through different angles on refraction. This accounts for the bands of colors produced by prisms, for example.

34–8 Total Internal Reflection

☐ When **total internal reflection** occurs, incident light is only reflected; it is not refracted into the neighboring region. This can occur when light is incident from a medium with a large index of refraction on a medium with a smaller index of refraction *and* the angle of incidence is greater than a certain critical value denoted by θ_c. Suppose $n_1 > n_2$ and light is incident from medium 1 on the boundary with medium 2. Then, $\theta_1 = \theta_c$ if $\theta_2 = 90°$ ($\sin \theta_2 = 1$). The critical angle is given by $\theta_c = \sin^{-1}(n_2/n_1)$.

34–9 Polarization by Reflection

☐ If unpolarized light is incident on a boundary between two different materials, both the reflected and refracted waves are partially polarized. For a special angle of incidence, called **Brewster's angle** and denoted by θ_B, the reflected wave is completely polarized. If θ_r is the angle of refraction for incidence at Brewster's angle, then $\theta_B + \theta_r = 90°$. If n_1 is the index of refraction for the medium of incidence and n_2 is the index of refraction for the medium of refraction, then Snell's law leads to $\tan \theta_B = n_2/n_1$. For the special case $n_1 = 1$ (air or vacuum), $\tan \theta_B = n_2$. This is **Brewster's law** for the polarizing angle.

☐ For incidence at Brewster's angle, the reflected light is polarized perpendicularly to the plane determined by the incident and reflected rays. Even at Brewster's angle, the *refracted* light contains components with all polarization directions, but is deficient in components with the polarization of the reflected light.

NOTES:

Chapter 35
IMAGES

The law of reflection is applied to plane and spherical mirrors and the law of refraction is applied to spherical refracting surfaces. Each of these form an image of an object placed in front of it and you will learn to find the position, size, and orientation of the image. You will also apply what you learn to lenses with spherical faces and to systems of lenses, such as those used in telescopes and microscopes. Pay careful attention to ray tracing techniques, which help you to visualize image formation.

Important Concepts

☐ plane mirror
☐ real image
☐ virtual image
☐ object distance
☐ image distance
☐ spherical mirror
☐ focal length of a mirror

☐ focal point of a mirror
☐ lateral magnification
☐ spherical refracting surface
☐ lens
☐ focal length of a lens
☐ focal points of a lens

35–1 Two Types of Image

☐ Light rays diverge from an image and your brain thinks the image is at the place from which the rays diverge. An image is said to be a **real image** if light actually passes through the image point. It is said to be a **virtual image** if light does not. The image in a plane mirror, for example, appears to come from behind the mirror but it actually diverges from the surface of the mirror. The image is virtual.

35–2 Plane Mirrors

☐ After light from an object has been reflected by a plane mirror, it appears to come from an **image** located behind the mirror. The diagram shows a point source (or **object**) P in front of a mirror and two rays emanating from it. The reflected rays are drawn according to the law of reflection. The dotted lines extend the reflected rays to behind the mirror, where they intersect at the image point P'.

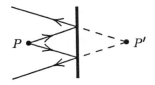

Many other rays from the source to the mirror could have been drawn. All reflected rays, extended backward to the region behind the mirror, intersect at P'. P' is on the line through P that is perpendicular to the mirror and is the same distance behind the mirror as P is in front.

☐ If the source is extended, you may think of each point on it as a point source, emitting light in all directions. An image is formed for each point that sends light to the mirror. To find the image of a straight line, simply find the images of each end and connect them with a straight line. Clearly the image of an extended source is the same size as the source.

☐ The distance from a point source to a mirror is denoted by p. It is positive. The image position is denoted by i. It is positive for real images, negative for virtual images, and its magnitude is the distance from the image to the mirror. For a plane mirror, $i = -p$.

35–3 Spherical Mirrors

☐ The diagrams below show two spherical mirrors, with their centers labelled c and their centers of curvature labelled C. A point source of light S is in front of each mirror and its image is formed at I. F marks the focal point of the mirror. The line through the center of curvature and the center of the mirror is called the **central axis**. Angles are usually measured with respect to this line and distances are measured along it. The object distance is p and the image distance is i.

☐ Light from the source is reflected by the mirror. In some cases, reflected rays converge to a point on the source side of the mirror and form a real image. In other cases, the reflected rays diverge as they leave the mirror but they follow lines that pass through a single point behind the mirror. They form a virtual image. Strictly speaking, sharp images are formed only by light whose rays make small angles with the central axis.

☐ Incident rays that are parallel to the central axis of a *concave* mirror, like the mirror on the left above, are reflected so they pass through the focal point, which is a distance $f = r/2$ in front of the mirror. Incident rays that are parallel to the central axis of a *convex mirror*, like the mirror on the right above, are reflected so they diverge, but they appear to come from the focal point, a distance $r/2$ behind the mirror. Here r is the radius of curvature of the mirror.

☐ A mirror that is concave with respect to the source has a positive radius of curvature and focal length; a mirror that is convex with respect to the source has a negative radius of curvature and focal length.

35–4   Images from Spherical Mirrors

☐ The mirror equation relates the object distance p, the image distance i, and the radius of curvature r. It is

$$\frac{1}{p} + \frac{1}{i} = \frac{2}{r} .$$

This is often written in terms of the **focal length** of the mirror, defined by $f = r/2$:

$$\frac{1}{p} + \frac{1}{i} = \frac{1}{f} .$$

Both r and f are positive if the center of curvature is on the same side of the mirror as the object and negative if the center of curvature is on the opposite side. The image distance i is positive for a real image (on the same side of the mirror as the object) and negative for a virtual image (on the opposite side).

☐ **Ray tracing** provides a graphical way of finding the image of an object that is not on the central axis. Special rays are drawn from the object to the mirror, then the reflected rays are drawn. They intersect at the image. The special rays are:

1. An incident ray that is parallel to the central axis is reflected so the reflected ray, perhaps extended backward, is through the focal point.

2. An incident ray that passes through the focal point is reflected so the reflected ray is parallel to the central axis.

3. An incident ray that passes through the center of curvature is reflected so the reflected ray is along the same line.

The three special rays are illustrated below. If the image is real, it is at their intersection. If it is virtual, it is at the intersection of their extensions to the region behind the mirror.

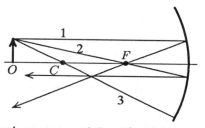

□ If the mirror is concave and the object is between the focal point and the mirror, then the image is virtual and erect. If the object is outside the focal point, it is real and inverted. If the object is more than twice the focal length, from the mirror the image is smaller than the object; otherwise it is larger. If the object is at the focal point, all small-angle reflected rays are parallel to each other and the image is said to be at infinity. If the mirror is convex, the image is virtual and erect, no matter where the object is. These statements can be verified easily by tracing rays or by using the mirror equation.

□ The **lateral magnification** m of a mirror is the ratio of the lateral size (the dimension normal to the central axis) of the image to the lateral size of the object. The lateral magnification of a spherical mirror is given by $m = -i/p$. Values for p and i are substituted with their signs. If m is positive, then the orientations of the object and image are the same; if m is negative, then the image is inverted with respect to the object. Virtual images of an erect object are erect; real images of an erect object are inverted.

35–5 Spherical Refracting Surfaces

□ Suppose a spherical surface separates the medium of incidence, with index of refraction n_1, from the medium of refraction, with index of refraction n_2. The central axis is the line through the center of the surface and its center of curvature. If rays from a point source on the central axis make small angles with that axis, then, after refraction, they either converge on a real image or else their extensions diverge from a virtual image. If the image is virtual, it is formed in the region containing the source. If the image is real, it is formed on the side into which light is transmitted.

☐ Distances are measured along the central axis from the surface. For a single refracting surface, the object distance p is positive. The image distance i is positive if the image is real and negative if it is virtual.

☐ The law of refraction yields a relationship between the object distance p, the image distance i, the radius of curvature r of the surface, and the indices of refraction for the two sides:

$$\frac{n_1}{p} + \frac{n_2}{i} = \frac{n_2 - n_1}{r} .$$

If the surface is concave with respect to the source, the center of curvature is on the same side of the surface as the object and r is negative. If the surface is convex with respect to the source, the center of curvature is on the opposite side and r is positive.

☐ If $n_2 > n_1$, then a convex surface forms a real image of an object that is far from it and a virtual image of an object that is near it. Details depend on the values of the indices of refraction and the radius of curvature. A concave surface, on the other hand, always forms a virtual image.

35–6 Thin Lenses

☐ A **lens** has two refracting surfaces. The image formed by the first is considered the object for the second. If the surrounding medium is a vacuum and the thickness of the lens can be neglected, then the object distance p, image distance i, and focal length f are related by

$$\frac{1}{p} + \frac{1}{i} = \frac{1}{f} .$$

The focal length f is given by

$$\frac{1}{f} = (n-1)\left(\frac{1}{r_1} - \frac{1}{r_2}\right).$$

Here r_1 is the radius of curvature of the surface nearer the object, r_2 is the radius of curvature of the surface farther away, and n is the index of refraction of the lens material.

☐ Real images are formed on the opposite side of the lens from the object. For these images, i is positive. Virtual images are formed on the same side of the lens as the object. For a virtual image, i is negative. A surface radius is negative if the center of curvature is on the same side of the surface as the object and is positive if the center of curvature is on the opposite side. The focal length of a lens does not depend on which side faces the object. Lenses with positive focal lengths are said to be converging, while lenses with negative focal lengths are said to be diverging.

☐ A lens has two focal points, located equal distances $|f|$ on opposite sides of the lens. The first focal point, denoted F_1, is on the same side of the lens as the object for a converging lens (f positive) and on the opposite side for a diverging lens (f negative). Light rays that are along lines that pass through F_1 are bent by the lens to become parallel to the axis. For a converging lens, the rays before refraction actually pass through F_1. For a diverging lens, the rays strike the surface, so only their extensions into the other side pass through F_1.

☐ The second focal point, denoted by F_2, is on the side opposite the object for a converging lens (f positive) and on the same side as the object for a diverging lens (f negative). Light rays that are parallel to the central axis are bent by the lens to lie along lines that pass through F_2. For a converging lens, the rays after refraction actually do pass through F_2. For a diverging lens the, backward extensions of the rays pass through F_2.

☐ The position of an off-axis image can be found graphically by tracing two or more rays originating at the same point on the object. One of these might be along a line through F_1 and then parallel to the axis; another might be parallel to the axis and then along a line through F_2. A third ray you might use goes through the center of the lens. It is not refracted. The image is at the point of intersection of these rays or their backward extensions.

☐ In terms of the object and image distances, the lateral magnification associated with a lens is given by $m = -i/p$, an expression that is identical to the expression for a spherical mirror. If m is negative, the image of an erect object is inverted; if m is positive, the image of an erect object is erect. A virtual image formed by a single thin lens is always erect; a real image is always inverted.

35–7 Optical Instruments

☐ Because it takes into account the apparent diminishing of the size of an object with distance, the **angular magnification** is often a better measure of the usefulness of a lens used for viewing than is the lateral magnification.

☐ If an object has a lateral dimension h and is a distance d from the eye, its angular size is given in radians by $\theta = h/d$. If the object is viewed through a lens and the image is a distance d' from the eye and has a lateral dimension of h', then the angular size of the image is $\theta' = h'/d'$. The angular magnification of the lens is the ratio of these, or $m_\theta = \theta'/\theta = h'd/hd'$.

☐ A single converging lens used as a magnifying glass is usually positioned so the object is just inside the focal point F_1. The image is then virtual and far away. Its lateral size is $h' = mh = -ih/p$, where m is the lateral magnification. Thus, $m_\theta = -(i/p)(d/d')$. The eye is close to the lens so $d' \approx |i|$. Furthermore, $p \approx f$. Once these substitutions are made, the result is $m_\theta = d/f$. Usually d is taken to be the distance to the near point of the eye, about 25 cm, so $m_\theta = (25\,\text{cm})/f$.

☐ The near point is used since the object is in focus and has its largest angular size when it is this distance from an unaided eye. The angular magnification then tells us how much better the lens is than the best the unaided eye can do.

☐ Optical instruments, such as telescopes and microscopes, consist of a series of lenses. They can be analyzed by tracing a few rays as they pass through each lens in succession or by applying the lens equation to each lens in succession. Consider a system of two lenses, a distance ℓ apart on the same central axis. Let p_1 be the distance from the object to the first lens struck. This lens forms an image at i_1, given by $1/p_1 + 1/i_1 = 1/f_1$, where f_1 is the focal length of the lens. This image is the object for the second lens; the object distance is $p_2 = \ell - i_1$ and the image distance i_2 is given by $1/p_2 + 1/i_2 = 1/f_2$, where f_2 is the focal length of the second lens.

☐ For some systems, the object distance for the second lens may be negative. This occurs if the image formed by the first lens is behind the second lens, so the light exiting the first lens is converging as it strikes the second lens. Such objects are called *virtual* objects. The equation relating object and image distances is still valid — just substitute a negative value for p_2.

☐ A simple **microscope** consists of an objective lens and an eyepiece (or ocular). The focal length of the objective lens is small and the lens is positioned so the object lies just outside its first focal point. The image produced by the objective is real, large, inverted, and far from the lens. If the object has a lateral dimension h, then the image has a lateral dimension $h' = mh = ih/p$. The magnification is great since i is large and p is small.

☐ The distance s between the second focal point of the objective lens and the first focal point of the eyepiece is called the **tube length** of the microscope. The magnification of the objective lens is given by $m = h'/h = s/f_{ob}$.

☐ The eyepiece is positioned with its first focal point at the image produced by the objective lens. The eye is placed close to the eyepiece, which then acts as a simple magnifying glass with an angular magnification of $(25\text{ cm})/f_{ey}$, where f_{ey} is the focal length of the eyepiece. The overall magnification is given by $M = mm_{\theta} = -(s/f_{ob})(25\text{ cm}/f_{ey}$. The expression compares the angular size of the image produced by the microscope with the angular size of the object when it is at the near point of an unaided eye.

☐ Telescopes are used to view objects that are far away. Rays entering the objective lens are essentially parallel to the central axis and that lens produces an image that is close to its second focal point. To compute the angular magnification of a telescope, we compare the angular size of the object at its far-away position (not the near point) to the angular size of the image produced by the telescope. The result is $m_{\theta} = -f_{ob}/f_{ey}$. To obtain a large angular magnification, a telescope should have an objective lens with a long focal length and an eyepiece with a short focal length.

NOTES:

Chapter 36
INTERFERENCE

You studied the fundamentals of interference in Chapter 17. Now the results are specialized to light waves and applied to double-slit interference, thin-film interference, and the Michelson interferometer, an important instrument for measuring distances. Pay special attention to the role played by the distances traveled by interfering waves in determining their relative phase. Also remember that there may be a phase change on reflection.

Important Concepts

☐ interference   ☐ coherence

☐ Huygens' principle   ☐ thin-film interference

☐ diffraction   ☐ phase change on reflection

☐ double-slit interference pattern   ☐ Michelson interferometer

---

36–1 Interference

☐ The electric fields of two electromagnetic waves at the same place add vectorially. If the fields are in the same direction, the resultant wave has a larger amplitude than either of the constituents; if they are in opposite directions, the resultant wave has a smaller amplitude. Completely destructive interference, in which the fields cancel each other, may occur. Clearly, the wave nature of electromagnetic radiation is important for interference phenomena.

36–2 Light as a Wave

☐ Maxwell's equations lead to Huygens' principle, a geometrical construction that shows how to construct the wavefront for an electromagnetic wave at some time from the wavefront at an earlier time. According to the principle, each point on a wavefront acts as a point source of spherical waves, called Huygens wavelets. After time Δt, the radius of the wavelets will be $v\Delta t$, where v is the wave speed, and the wavefront will be tangent to the wavelets.

☐ The law of refraction follows directly from Huygens' principle. If a wavefront in the incident medium is not parallel to the refracting surface, then part of it enters the second medium before the rest. If the index of refraction of the second medium is greater than the index of refraction of the first, then the wavelets travel slower in the second medium and the wavefront turns to become more nearly parallel to the surface. The argument can be made quantitative and results in the law of refraction.

☐ When light with wavelength λ in vacuum propagates from a medium with index of refraction n_1 into a medium with index of refraction n_2, its wavelength changes from λ/n_1 to λ/n_2. Its frequency does not change.

36–3 Diffraction

☐ When a wave is incident on a hole in a barrier, the wave flares out into the region beyond the hole. This is **diffraction**. Huygens' principle explains the phenomenon as the propagation of wavelets into the region, emanating from the portion of the incident wavefront that is in the hole. The more narrow the hole, the greater is the spread of the waves on the other side. If the hole is wide (compared to a wavelength), there is little spreading.

36–4 Young's Interference Experiment

☐ In **Young's experiment**, a monochromatic plane wave is incident normally on a barrier with two slits and the intensity is observed on a viewing screen behind the slits. Wavelets arrive at every point on the screen from both slits and interfere to produce an intensity pattern there. The electric fields of the two waves are vectors and, strictly speaking, the fields should be added vectorially. Assume, however, that the waves are plane polarized and the fields are along the same line. Then, scalar addition can be used. This is a good approximation for most of the situations considered in this chapter.

☐ Assume the slits are so narrow that only one wavelet from each slit is required. Suppose the wavelet from one slit travels the distance r_1 to get to some point on the screen and the wavelet from the other slit travels the distance r_2 to get to the same point. At that point on the screen, the phase of the wavelet from the first slit is $\omega t - k r_1$ and the phase of the wavelet from the other slit is $\omega t - k r_2$, where $k = 2\pi/\lambda$ and λ is the wavelength. The phase difference is $\phi = k(r_2 - r_1) = 2\pi(r_2 - r_1)/\lambda$. A maximum in the intensity occurs if ϕ is a multiple of 2π rad; then, the wave crests of the two wavelets arrive at the screen at the same time. A minimum in the intensity occurs if ϕ is an odd multiple of π rad; then, the crests of one wavelet arrive at the same time as the troughs of the other.

☐ If the screen is far from the slits, then rays from the slits to any point on the screen are nearly parallel to each other and the difference in the distances can easily be written in terms of the angle θ between a ray and a line normal to the barrier. The geometry is shown in the diagram below. Since the screen is far away, the distance from the dotted line between the rays to a point on a screen is the same along each ray. The lower ray is longer than the upper by $d \sin \theta$, where d is the slit separation. The phase difference at the screen is $\phi = (2\pi d/\lambda) \sin \theta$.

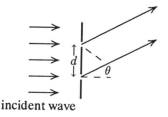

incident wave

□ The condition for a maximum is $(2\pi d/\lambda)\sin\theta = 2\pi m$, or $d\sin\theta = m\lambda$, where m is an integer. The condition for a minimum is $(2\pi d/\lambda)\sin\theta = (2m + 1)\pi$, or $d\sin\theta = (m + \frac{1}{2})\lambda$, where m is an integer.

□ Light emanates from each slit in *all* forward directions but only the portions of wavefronts that follow the rays at the angle θ get to the point on the screen being considered. Other portions of the wavefronts get to other places on the screen and for them θ has a different value. For different values of θ, the phase difference is different and, as a result, so are the resultant amplitude and intensity. Alternating bright and dark regions (fringes) are seen on the screen. Centers of bright fringes (maxima of intensity) occur at points for which $d\sin\theta = m\lambda$; centers of dark fringes (minima of intensity) occur at points for which $d\sin\theta = (m + \frac{1}{2})\lambda$.

□ Constructive interference occurs at the point on the screen directly in back of the slits, for which $\theta = 0$. The angular separation of the first minima on either side of the central maximum is a measure of the extent to which the intensity pattern is spread on the screen. This is given by $2\theta_0$, where $\sin\theta_0 = \lambda/2d$. As d decreases or λ increases, the angular separation increases and the pattern spreads. If $d = \lambda/2$, then $\theta_0 = 90°$ and no bright fringes appear beyond the central maximum.

36–5 Coherence

□ Two sinusoidal waves are **coherent** if the difference in their phases is constant. Light is emitted from atoms in bursts lasting on the order of nanoseconds and each burst may have a different phase constant associated with it. Thus, light from atoms emitting independently of each other is not coherent.

□ If an extended incoherent source, such as an incandescent lamp, is used to produce light that is incident on a double-slit barrier, the light from each atom goes through each slit and combines on the other side to form an interference pattern. Light from different atoms, however, form patterns that are shifted with respect to each other, the amount of the shift depending on the separation of the atoms in the source. The intensity at any point on the screen fluctuates rapidly between maximum and minimum. Since the eye cannot respond to the rapid fluctuations, a nearly uniform intensity is seen.

☐ A single-slit barrier, placed between the light source and the double-slit barrier, ensures that light from only a small region of the source passes through the double-slit barrier and that the intensity patterns produced at the screen by light from two different atoms nearly coincide. All the light from a laser is coherent, even though many different atoms are emitting simultaneously. When this light is incident on a double slit, it produces an interference pattern without additional apparatus.

36–6 Intensity in Double-Slit Interference

☐ Phasors can be used to sum waves. A phasor is a rotating arrow with length equal to the amplitude of the wave and angular velocity equal to the angular frequency of the wave. If a phasor has length E_0 and rotates counterclockwise with angular velocity ω, then its projection on the vertical axis is $E(t) = E_0 \sin(\omega t)$.

☐ The phasors for the electric fields of the wavelets from the two slits are drawn below. The waves are assumed to have the same amplitude E_0 and their phase difference is ϕ. The phasor for the resultant amplitude E is also drawn.

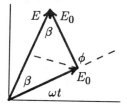

Geometry can be used to show that $\beta = \phi/2$ and that $E = 2E_0 \cos \beta = 2E_0 \cos(\phi/2)$.

☐ The intensity is proportional to the square of the amplitude, so

$$I = 4I_0 \cos^2(\tfrac{1}{2}\phi),$$

where $\phi = (2\pi d/\lambda) \sin \theta$ and I_0 is the intensity of a single wave. Maximum intensity occurs if $\cos(\phi/2) = \pm 1$; then, $\phi = 2m\pi$, $E = 2E_0$, and $I = 4I_0$. Minimum intensity occurs if $\cos(\phi/2) = 0$; then, $\phi = (2m + 1)\pi$, $E = 0$, and $I = 0$.

36–7 Interference from Thin films

☐ If light is incident normally on a thin film, some is reflected from the front surface and some from the back. After reflection, the two waves interfere and the resultant intensity may be large or small, depending on their phases and, hence, on the thickness of the film.

☐ For normal incidence on a film of thickness L and index of refraction n, the wave reflected from the back surface travels a distance $2L$ further than the wave reflected from the front surface. The wavelength in the film is $\lambda' = \lambda/n$, so the difference in path length produces a difference of $4\pi n L/\lambda$ in the phase of the two waves.

☐ For one or both of the waves, the medium beyond the reflecting surface might have a higher index of refraction than the medium of incidence. If it does, the wave suffers a phase change of π rad on reflection. Suppose the light is incident in a medium with index of refraction n_1, the film has index of refraction n_2, and the medium beyond the film has index of refraction n_3. Then, the phase difference for waves reflected from the two surfaces is $\phi = 4\pi n_2 L/\lambda$ if $n_1 > n_2 > n_3$ or $n_1 < n_2 < n_3$. In the first case, neither wave suffers a phase change on reflection. In the second, they both suffer a phase change of π rad. If $n_1 < n_2 > n_3$ or $n_1 > n_2 < n_3$, the phase difference is $\phi = (4\pi n_2 L/\lambda) \pm \pi$.

In the first case, the wave reflected from the front surface suffers a phase change of π rad but the wave reflected from the back surface does not. In the second case, the wave reflected from the back surface suffers a phase change of π rad but the wave reflected from the front surface does not. It is immaterial whether the sign in front of π is plus or minus.

□ If the phase difference ϕ is a multiple of 2π rad, the waves interfere constructively to produce a bright reflection. If ϕ is an odd multiple of π rad, the interference is completely destructive and the light is transmitted through the film; it is not reflected.

□ When white light (a combination of all wavelengths of the visible spectrum) is incident on a thin film, the reflected light is colored. It consists chiefly of those wavelengths for which interference of the two reflected waves produces a maximum or nearly a maximum of intensity. Wavelengths for which interference produces a minimum are missing. This phenomena accounts for the colors of oil films and soap bubbles, for example.

□ When monochromatic light is incident on a thin film with a varying thickness, like a wedge, interference produces bright and dark bands. Bright bands appear in regions for which the film thickness is such that constructive interference occurs; dark bands appear in regions for which the film thickness is such that destructive interference occurs.

36–8 Michelson's Interferometer

□ A schematic of the instrument is shown below. Light from a source S is incident on a half-silvered mirror M, where the beam is split. One beam travels to mirror M_2, back to M, and then to the eye at E. The other beam travels to mirror M_1, back to M, and then to E. The two beams reaching the eye interfere. To measure a distance, one of the mirrors M_1 or M_2 is moved, thereby changing the interference pattern at the eye. The change in the interference pattern is directly related to the distance moved by the mirror.

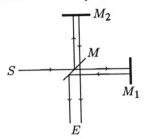

☐ Suppose an intensity maximum is produced at the eye, then either M_1 or M_2 is moved so a neighboring intensity minimum is produced. This means the mirror moved half a wavelength.

NOTES:

Chapter 37
DIFFRACTION

Diffraction is the flaring out of light as it passes by the edge of an object or through an opening in a barrier. Pay attention to the description of this phenomenon in terms of Huygens wavelets and learn how the interference of the wavelets produces bright and dark fringes in a diffraction pattern. Single- and double-slit patterns are considered in detail. Pay attention to the slit characteristics that determine the maxima and minima of each pattern. Two important applications are discussed. Diffraction patterns produced by diffraction gratings are used to study spectra and patterns produced by crystals are used to study crystalline structure. Your goals should be to learn what the patterns look like and to understand the details of wave interference that lead to these patterns.

Important Concepts

☐ single-slit diffraction
☐ double-slit diffraction
☐ circular aperture diffraction
☐ Rayleigh criterion
☐ multiple-slit diffraction

☐ diffraction grating
☐ dispersion
☐ resolving power
☐ x-ray diffraction
☐ Bragg's law

37–1   Diffraction and the Wave Theory of Light

☐ Diffraction is discussed in terms of Huygens wavelets. If there is no barrier, the wavelets combine to produce a wave that continues moving in its original direction. If a barrier blocks some of the wavelets, then those that are not blocked move into the geometric shadow. In addition, if the light is coherent, the Huygens wavelets interfere to form a series of bright and dark bands, called the diffraction pattern of the object. Figs. 37–1, 37–2, and 37–3 of the text show some diffraction patterns.

37–2 Diffraction By a Single Slit: Locating the Minima

☐ Consider plane waves of monochromatic light incident normally on a barrier with a single slit of width a, as shown below. To find the intensity at a point P on a screen, add the Huygens wavelets emanating from the slit and take the limit as the number of wavelets becomes infinite. Suppose the viewing screen is far away from the slit and consider parallel rays.

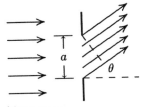

incident wave

☐ The region of screen directly behind the slit is bright. This is the central maximum. The first minimum of the diffraction pattern occurs when every wavelet from the upper half of the slit can be paired with a wavelet from the lower half and the two wavelets of a pair are π rad out of phase. The distance between the two source points for a pair of wavelets is $a/2$, where a is the width of the slit, and the difference in the distance traveled by the wavelets is $(a/2)\sin\theta$. Thus, the condition for the first minimum is $(a/2)\sin\theta = \lambda/2$, or $a\sin\theta = \lambda$. Other minima occur for $a\sin\theta = m\lambda$, where m is an integer.

☐ As a decreases, the angle θ for the first minimum increases. The central maximum is broader for narrow slits than for wide slits and diffraction is more pronounced. If a is less than λ, no zeros of intensity occur ($\sin\theta$ cannot be greater than 1) and the entire screen is within the central maximum.

37–3 Intensity in Single-Slit Diffraction, Qualitatively

☐ For $\theta = 0$, all the wavelets are in phase at the screen and they produce the bright central fringe. For other directions, the phases of wavelets from neighboring points in the slit differ and the phasors form the arc of a circle, as shown below. The amplitude of the resultant wave and the intensity are less than for $\theta = 0$. For larger θ, the arc covers a greater portion of a circle, so the amplitude and intensity are still less. At the first minimum, the phasors form a complete circle and the amplitude of the resultant is zero. For θ beyond the first minimum, the phasors wrap around a circle more than once. Since the total arc length is the same, the amplitude and intensity at a secondary maximum are less than at the central maximum.

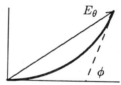

37–4 Intensity in Single-Slit Diffraction, Quantitatively

☐ The angle ϕ in the diagram above is the difference in phase of wavelets from the upper and lower edges of the slit. It is also the angle subtended by the arc at its center. If the arc has radius R, then its length is given by $E_m = R\phi$, for ϕ in radians. E_m is the sum of the amplitudes of all the wavelets and, thus, is the amplitude if they all have the same phase. E_θ, the amplitude at the screen, is the chord of the arc. A little geometry shows that $E_\theta = 2R\sin(\phi/2)$. Eliminating R between these two expressions yields

$$E_\theta = \frac{2E_m \sin(\phi/2)}{\phi},$$

where $\phi = (2\pi a/\lambda)\sin\theta$. The expression for the amplitude is often written in terms of $\alpha = \phi/2$, rather than in terms of ϕ. Then, the amplitude at the screen is

$$E_\theta = \frac{E_m \sin \alpha}{\alpha}$$

and the intensity is

$$I = I_m \left(\frac{\sin \alpha}{\alpha} \right)^2 ,$$

where I_m is the intensity at the center of the central fringe and $\alpha = (\pi a / \lambda) \sin \theta$.

☐ Carefully study Fig. 37–7 of the text. It shows the intensity as a function of θ for several values of the slit width a. The most prominent feature is a broad central maximum, centered at $\theta = 0$. Since, in the limit as $\alpha \to 0$, $(\sin \alpha)/\alpha \to 1$, the intensity equation predicts the maximum. If the slit width is small, the central maximum spreads to cover the entire screen and no zeros of intensity occur. For a wide slit, the central maximum is narrow and is followed on both sides by secondary maxima. These are narrower and of considerably less intensity than the central maximum. They are roughly midway between zeros of intensity. The number that appears depends on the slit width.

37–5 Diffraction By a Circular Aperture

☐ When plane waves pass through a circular aperture and onto a screen, a diffraction pattern is formed there. The pattern consists of a bright central disk, followed by a series of alternating dark and bright rings. The first minimum occurs at an angle θ that is given by $\sin \theta = 1.22\lambda/d$, where d is the diameter of the aperture and θ is measured from the normal to the aperture. The angle for the first dark ring can be used as a measure of the angular size of the central disk.

☐ Stars are effectively point sources of light and lenses act like circular apertures. The image of a star formed by a lens is broadened by diffraction from a point to a disk and rings. Two stars do not form distinct images if the central disks of their diffraction patterns overlap too much. According to the Rayleigh criterion, two far-away point sources are resolved if the centers of their diffraction patterns are no closer than the radius of the first dark ring.

37–6 Diffraction By a Double Slit

☐ Consider two identical slits, each of width a, with a center-to-center separation d. Any point on a screen is reached by a wave from each slit, the resultant of the Huygens wavelets from that slit, and we may think of the pattern as being formed by the interference of these two waves. If light is incident normally on the slits and the observation point is far away, the wave that reaches it from each slit has amplitude $E_m (\sin \alpha)/\alpha$, where $\alpha = (\pi a/\lambda) \sin \theta$, and the two waves differ in phase by $(2\pi d/\lambda) \sin \theta$. When they are combined, the resultant amplitude is

$$E_\theta = E_m (\cos \beta)^2 \left(\frac{\sin \alpha}{\alpha} \right)^2 ,$$

where $\beta = (\pi d/\lambda) \sin \theta$.

☐ Minima of the double-slit interference pattern occur for $\beta = (2m + 1)\pi/2$ or $\sin \theta = (2m + 1)\lambda/2d$. Single-slit diffraction minima occur for $\alpha = m\pi$ or $\sin \theta = m\lambda/a$. In each case, m is an integer but it may have different values in the two expressions. Since d must be larger than a, the interference minima must be closer together than the diffraction minima. The single-slit diffraction pattern forms an envelope, with the interference pattern inside. Study Fig. 37–14 of the text.

☐ Three parameters are important for the double-slit pattern: the wavelength λ, the slit width a, and the center-to-center slit separation d. The ratio a/λ determines the width of the central diffraction maximum, which extends from $\theta = -\sin^{-1}(\lambda/a)$ to $\theta = +\sin^{-1}(\lambda/a)$. This ratio also determines the positions of the secondary diffraction maxima, if any. The ratio d/λ controls the angular positions of the interference maxima and minima. Finally, the ratio d/a determines how many interference maxima fit within the central diffraction maximum or any of the secondary diffraction maxima.

37–7   Diffraction Gratings

☐ A diffraction grating consists of many thousands of closely spaced rulings on either a transparent or highly reflecting surface. When light is incident on a grating, a multiple-slit diffraction pattern is formed. Because light with different wavelengths produces lines at different angles, diffraction gratings are often used to analyze the spectra of light sources.

☐ Consider a barrier with N parallel slits. Monochromatic plane waves are incident normally on the barrier and the intensity pattern formed by waves passing through the slits is viewed on a screen far away. If the slits are narrow, single-slit diffraction can be ignored. That is, the interference pattern is well within the central maximum of the single-slit diffraction pattern.

☐ The pattern on the screen consists of a series of intense, narrow bands, called **lines**. Secondary maxima lie between but they are much less intense and are usually not important. The pattern is usually described in terms of the angle θ made with the normal by a ray from the slit system to a point on the screen.

☐ A diffraction line occurs when the phases at the screen of waves from any two adjacent slits are either the same or differ by a multiple of 2π rad. Since waves from two adjacent slits travel distances that differ by $d\sin\theta$, where d is the slit separation, the condition for a line is $d\sin\theta = m\lambda$, where λ is the wavelength. The integer m in this equation is called the **order** of the line. High order lines occur at greater angles than low order lines.

☐ The locations of the lines are determined by the ratio d/λ and are independent of the number of slits. The lines occur at different angles for different wavelengths of light. For the same order line, the angle for red light is smaller than that for violet light. If white light is incident on the barrier, the color of an observed band continuously varies from red at one end to violet at the other.

☐ The width of a line is indicated by the angular separation $\Delta\theta$ of the minima on either side. For the line that occurs at angle θ, $\Delta\theta = \lambda/Nd\cos\theta$. It depends on the number of slits. In fact, as the number of slits increases without change in their separation, the width of every line decreases. Maxima near the normal (small θ) are sharper than maxima away from the normal (larger θ).

37–8 Gratings: Dispersion and Resolving Power

☐ Two parameters are used to measure the quality of a grating. The **dispersion** measures the angular separation of lines of the same order for wavelengths differing by $\Delta\lambda$. It is defined by $D = \Delta\theta/\Delta\lambda$ and for order m, occurring at angle θ, it is given by $D = m/d\cos\theta$, where d is the slit separation. Dispersion does not depend on the number of rulings. Large dispersion means large angular separation.

☐ The **resolving power** measures the difference in wavelength for lines with an angular separation equal to half their angular width; that is, for two lines that obey the Raleigh criterion for resolution. Mathematically, it is defined by $R = \lambda/\Delta\lambda$, where $\Delta\lambda$ is the difference in wavelength. For an N-ruling grating, the resolving power for order m is given by $R = Nm$.

37–9 X-Ray Diffraction

☐ Atoms in a crystal form a periodic array in three dimensions and x-ray radiation scattered by their electrons produces a diffraction pattern. To obtain a measurable scattering angle, the wavelength of the radiation used should be about the same as the distance between atoms. Thus, x-ray radiation, with a wavelength on the order of 0.1 nm, is used.

☐ Diffraction occurs only when the x rays are incident at certain angles to **crystal planes**. A crystal plane is a plane that passes through atomic equilibrium positions.

☐ Suppose monochromatic x rays with wavelength λ are incident at an angle θ on a set of crystal planes with separation d. If the angle θ between the beam and the planes satisfies $2d \sin \theta = m\lambda$, where m is an integer, then a high intensity beam is radiated at the angle θ to the planes. This is Bragg's law for x-ray diffraction.

☐ X-ray diffraction is used to find the orientations and separations of a great many sets of crystal planes. These are used to reconstruct the crystal. Crystals with known atomic arrangements are used as filters to separate x rays of a given wavelength from an incident beam containing a mixture of wavelengths.

NOTES:

Chapter 38
RELATIVITY

When two observers who are moving relative to each other measure the same physical quantity, they may obtain different values. The special theory of relativity tells how the values are related to each other when both observers are at rest in different inertial frames. Although the complete theory deals with all physical quantities, the ones you consider here are the coordinates and time of an event and the velocity, momentum, and energy of a particle.

Important Concepts

- [ ] special theory of relativity
- [ ] event
- [ ] simultaneity
- [ ] time dilation
- [ ] proper time
- [ ] Lorentz factor
- [ ] length contraction

- [ ] proper length
- [ ] Lorentz transformation
- [ ] velocity transformation
- [ ] relativistic Doppler effect
- [ ] relativistic momentum
- [ ] relativistic kinetic energy
- [ ] rest energy

38–2 The Postulates

- [ ] The **special theory of relativity** is based on two postulates. The first (the relativity postulate) deals with the laws of physics: The laws of physics are the same for observers in all inertial frames. Keep in mind that the laws of physics are relationships between physical quantities, not the quantities themselves. Newton's second law and the conservation principles are examples of laws. The momentum of a system has a different value for different reference frames but if it is conserved for one inertial frame, it is conserved for all inertial frames, according to the postulate.

☐ The second postulate (the speed of light postulate) is: The speed of light in free space has the same value in all directions and in all inertial frames. If a light source sends a pulse of light toward you, the speed of the pulse, as observed by you, is the same if you are at rest relative to the source, you are traveling at high speed toward (or away from) the source, or the source is traveling at high speed toward (or away from) you. This postulate is consistent with the notion that electromagnetic radiation does not require a medium for its propagation.

☐ Special relativity deals with measurements made in *inertial reference frames*. If the total force on a particle is zero, then its acceleration, as measured in an inertial frame, is zero.

38–3 Measuring an Event

☐ An **event** has four numbers associated with it: three are coordinates that designate the location of the event; the fourth designates the time of the event. To measure the coordinates of an event, meter sticks must be laid out in an inertial reference frame, at rest with respect to the frame. To measure the time of an event, a clock must be present at the location of the event and it must be at rest with respect to the reference frame. In addition, it must be synchronized with other clocks at rest in the frame. Synchronization is accomplished, for example, by sending a light pulse from a master clock to all other clocks and using its arrival time at any clock to set that clock, taking into account the transit time of the pulse.

☐ The coordinates and time of any event may be measured by meter sticks and clocks at rest in any inertial frame. Relativity theory tells how the measurements made in one frame are related to those made in another.

38–4 The Relativity of Simultaneity

☐ Two events, separated in space and simultaneous to one observer, are NOT simultaneous to another observer, moving with respect to the first along the line joining the positions of the events. Carefully study Fig. 38–4 of the text. It shows two events, labeled Red and Blue, that are simultaneous according to Sam. He knows they are simultaneous because the events occurred at the ends of his spaceship and electromagnetic waves from the events meet at the midpoint. Since the waves travel at the same speed and go the same distance, they must have started at the same time.

The events also occur at the ends of Sally's space ship but she is moving away from the position of the Blue event and toward to the position of the Red event. Waves from the Red event reach the midpoint of her spaceship before waves from the Blue event. In her frame, the waves move with the speed of light, just as they do in Sam's frame, so she knows that the Red event occurred before the Blue event.

☐ The situation is exactly symmetric. If two events are simultaneous in Sally's frame, they are not simultaneous in Sam's.

38–5   The Relativity of Time

☐ The interval of time between two events is different when measured with clocks at rest in two inertial frames that are moving relative to each other. The digram on the left below shows a clock. The flash unit F emits a light pulse that travels to mirror M and is reflected back to the flash unit. It is detected there and immediately triggers the next flash. If the flash unit and mirror are separated by a distance D, then the time interval between flashes is given by $\Delta t_0 = 2D/c$.

The diagram on the right below shows what Sam sees when Sally carries a clock like this at speed v past him. If Δt is the time interval between emission and detection of the pulse, as measured by Sam, then during this interval the flash unit travels a distance $\ell = v \Delta t$ and the light pulse moves a distance $2L = 2\sqrt{(\frac{1}{2}v \Delta t)^2 + D^2}$. This follows because L is the hypotenuse of a right triangle with sides of length D and $v\,\Delta t/2$. Substitute $2L = c\Delta t$ and solve for Δt:

$$\Delta t = \frac{2D}{\sqrt{c^2 - v^2}} = \frac{\Delta t_0}{\sqrt{1 - (v/c)^2}}.$$

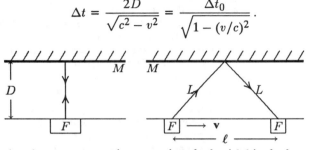

An observer comparing a moving clock with his clocks concludes that the moving clock ticks at a slower rate. This phenomenon is called **time dilation**. It is not important that the clock utilize light as does the clock used to derive the relationship above. Any clock will do.

☐ The time dilation equation is often written

$$\Delta t = \gamma \Delta t_0,$$

where the **Lorentz factor** γ is $1/\sqrt{1 - (v/c)^2}$. The speed is often given as a fraction of the speed of light. In terms of β (= v/c), $\gamma = 1/\sqrt{1 - \beta^2}$. Since the speed of a reference frame is always less than the speed of light, the factor γ is always greater than one.

☐ Time dilation is symmetric with respect to the two frames. If Sally watches a clock at rest with respect to Sam, she sees it tick at a slower rate than her clocks.

☐ If two events, such as the emission and detection of a light pulse, occur at the same coordinate in one frame, then the time interval between them, as measured in that frame, is the **proper time** between the events. The time interval between the same two events, as measured in a frame that is moving relative to the first, is longer by the factor γ.

38–6   The Relativity of Length

☐ If you measure the length of a rod that is moving past you at high speed, the result is less than if you measure it when it is at rest with respect to you. The length measured with the rod at rest relative to the meter stick is called the **proper length** of the rod.

☐ Suppose the speed of the rod is v. Place a marker on a coordinate axis parallel to the rod's velocity and measure the interval from the time the front end of the rod is at the marker to the time the back end of the rod is at the marker. If that interval is Δt_0, then the length of the moving rod is $L = v\,\Delta t_0$. A single clock at the position of the marker can be used, so Δt_0 is the proper time interval between the two events. L is NOT the proper length because the rod is moving relative to the frame used to measure the length.

Now consider the same events from the point of view of someone moving with the rod. He sees the marker move with speed v from the front to the back of the rod in time Δt and gives the length of rod as $L_0 = v\,\Delta t$. This is the proper length because the rod is at rest relative to the observer. Since Δt_0 is the proper time interval between the events, Δt and Δt_0 are related by $\Delta t = \gamma\,\Delta t_0$, where $\gamma = 1/\sqrt{1 - (v/c)^2}$. Thus, in terms of its proper length, the length of the moving rod is

$$L = v\,\Delta t_0 = \frac{v\,\Delta t}{\gamma} = \frac{L_0}{\gamma}.$$

38–7   The Lorentz Transformation

☐ Suppose an event is viewed by two observers, one at rest in inertial frame S and another at rest in inertial frame S'. Also suppose that, from the viewpoint of S, S' is moving in the positive x direction with velocity v. The coordinate systems and clocks are arranged so that the origins coincide at time $t = 0$ and the clocks of S' are synchronized to read

0 when the origins coincide. The coordinates of the event are x, y, z and the time of the event is t, all as measured in S. The same event has coordinates x', y', z' and occurs at time t', as measured in S'. Then, according to the **Lorentz transformation**, the coordinates and times are related by

$$x' = \gamma(x - vt), \quad y' = y, \quad z' = z, \quad \text{and} \quad t' = \gamma(t - vx/c^2),$$

where $\gamma = 1/\sqrt{1 - (v/c)^2}$.

☐ Suppose the displacement of two events has components Δx, Δy, and Δz in S and has components $\Delta x'$, $\Delta y'$, and $\Delta z'$ in S'. The time interval between the events is Δt in S and $\Delta t'$ in S'. Then, the intervals are related by $\Delta x' = \gamma(\Delta x - v\,\Delta t)$, $\Delta y' = \Delta y$, $\Delta z' = \Delta z$, and $\Delta t' = \gamma(\Delta t - v\,\Delta x/c^2)$. The same equations can be used if S' is moving in the negative x direction relative to S. The velocity v is then negative.

☐ These equations can be solved for Δx, Δy, Δz, and Δt: $\Delta x = \gamma(\Delta x' + v\,\Delta t')$, $\Delta y = \Delta y'$, $\Delta z = \Delta z'$, and $\Delta t = \gamma(\Delta t' + v\,\Delta x'/c^2)$.

38–8 Some Consequences of the Lorentz Equations

☐ You can use the Lorentz transformation equations to investigate simultaneity. Suppose two events are simultaneous in S and are spatially separated by Δx. Set $\Delta t = 0$. The time interval between them, as measured in S', is given by $\Delta t' = -\gamma v\,\Delta x/c^2$, if S' is moving with speed v in the positive x direction relative to S. Now suppose two events are simultaneous in S' and are spatially separated by $\Delta x'$. Then, $\Delta t = \gamma v\,\Delta x'/c^2$.

☐ The Lorentz transformation equations predict time dilation. Take $\Delta x' = 0$ (the events occur at the same coordinate in S') and solve for Δt. You should obtain $\Delta t = \gamma \Delta t'$. The proper time interval between the events is measured in the primed frame. Now suppose the two events occur at the same coordinate in the S frame. Take $\Delta x = 0$ and solve for $\Delta t'$. You should obtain $\Delta t' = \gamma \Delta t$. The proper time interval between the events is now measured in the unprimed frame. Two events might occur at different coordinates in both frames. Then, the clocks in neither frame measure the proper time interval between them.

☐ The Lorentz transformation equations predict length contraction. Suppose an object of proper length L_0 is at rest along the x axis of frame S and its length is measured in S', a frame that is moving with speed v in positive x direction relative to S. One way of measuring the length is to place two marks on the x' axis of S', one at the front of the object and one at the back, then measure the distance between the marks. The marks, of course, must be made simultaneously. Set $\Delta x' = L$, $\Delta t' = 0$, and $\Delta x = L_0$. The Lorentz transformation equations yield $\Delta x' = \Delta x / \gamma$. Now suppose the object is at rest on the x' axis of S' and its length is measured in S. Set $\Delta x' = L_0$, $\Delta x = L$, and $\Delta t = 0$. The transformation equations now give $\Delta x = \Delta x' / \gamma$.

38–9 The Relativity of Velocities

☐ Suppose a particle is moving with velocity u' along the x' axis of S'. Then, its velocity as measured in S is given by

$$u = \frac{u' + v}{1 + u'v/c^2} \, ,$$

where v is the velocity of S' relative to S. This expression can be derived from the Lorentz transformation equations. Divide $\Delta x = \gamma(\Delta x' + v \, \Delta t')$ by $\Delta t = \gamma(\Delta t' + v \, \Delta x'/c^2)$ to obtain $\Delta x / \Delta t = (\Delta x' + v \, \Delta t')/(\Delta t' + v \, \Delta x'/c^2)$. Now divide both the numerator and denominator by $\Delta t'$, replace $\Delta x'/\Delta t'$ with u', and replace $\Delta x / \Delta t$ with u. The result is as given above.

☐ If $v \ll c$, the quantity $1 + u'v/c^2$ in the denominator can be approximated by 1 and the Galilean velocity transformation equation is obtained: $u = u' + v$.

☐ If the speed of an object, as measured in one frame, equals the speed of light, then its speed, as measured in any other frame, also equals the speed of light. No matter how fast you travel toward or away from an approaching light pulse, its speed relative to you is c. Prove this for an object moving in the positive x direction: replace u' with c in the expression for u and show that the result is $u = c$.

38–10   Doppler Effect for Light

☐ When a source of light is moving toward an observer or an observer is moving toward a source, the observed frequency is greater than the frequency f_0 measured in the rest frame of the source (the proper frequency). When the relative motion of the source and observer is one of increasing separation, the observed frequency is less than the proper frequency. Relativity predicts that the observed frequency is given by

$$f = f_0 \sqrt{\frac{1 - \beta}{1 + \beta}},$$

where $\beta = v/c$ and v is the relative velocity of the source and observer. If the distance between the source and observer is increasing, β is positive; if the distance is decreasing, β is negative.

☐ A Doppler shift also occurs if the motion is transverse to the line joining the source and observer. Then, the observed frequency is given by

$$f = f_0 \sqrt{1 - \beta^2}.$$

This expression is a direct result of the time dilation of the period of oscillation.

38–11 A New Look at Momentum

☐ Relativity theory requires a revision of the definition of momentum if it is to obey the familiar conservation law. More precisely, if measurements taken in one inertial frame show that momentum is conserved, then measurements taken in any other inertial frame should also show that momentum is conserved.

☐ If a particle has mass m and travels with velocity $\mathbf{v}$, then its momentum $\mathbf{p}$ is given by

$$\mathbf{p} = \gamma m \mathbf{v}.$$

If $v \ll c$, this expression reduces to $\mathbf{p} = m\mathbf{v}$, the non-relativistic definition.

☐ If a system consists of several particles, the total momentum is the vector sum of the individual momenta. If the net external force on the particles of a system vanishes, then the total momentum of the system is conserved.

38–12 A New Look at Energy

☐ Relativity theory also requires that the definition of kinetic energy be revised if it is to obey the familiar conservation law. If energy is conserved in one inertial frame, then it should be conserved in all inertial frames.

☐ The kinetic energy of a particle with mass m moving with speed v is given by

$$K = mc^2 \left(\frac{1}{\sqrt{1 - (v/c)^2}} - 1 \right).$$

This expression is valid no matter what the speed v. If $v \ll c$, it reduces to the familiar non-relativistic expression $K = \frac{1}{2}mv^2$. The total kinetic energy of a system of particles is the scalar sum of the individual kinetic energies.

This quantity is conserved in collisions provided the particle masses do not change. Work must be done to change the kinetic energy of a particle and the change in kinetic energy equals the net work done.Ł

☐ In many nuclear scattering and decay processes, the particle masses do change. You must then take into account the **rest energies** of the particles. The rest energy of a particle of mass m is given by $E_0 = mc^2$. The total energy of a particle, the sum of the rest and kinetic energies, is

$$E = mc^2 + K = \frac{mc^2}{\sqrt{1 - (v/c)^2}} \, .$$

☐ The total energy and the magnitude of the momentum of any particle are related by

$$E^2 = (pc)^2 + (mc^2)^2 \, .$$

This expression replaces $E = p^2/2m$, the non-relativistic relationship between E and p. The energy E in the relativistic relationship includes both rest and kinetic energies.

NOTES:

Chapter 39
PHOTONS AND MATTER WAVES

With this chapter you begin your study of quantum mechanics. You will learn that electromagnetic radiation, which is described classically as a wave-like propagation of electric and magnetic fields, also has particle-like properties. Learn to use conservation of energy and momentum to explain the experiments that provide evidence for the particle nature of electromagnetic radiation. You will learn that particles, such as electrons, also have waves associated with them. Matter waves give rise to interference and diffraction effects and to tunneling phenomena. Pay particular attention to the relationship between the energy of a particle or photon and the frequency of the wave and to the relationship between the momentum of the particle or photon and the wavelength of the wave. For both light and matter, pay attention to the interpretation of the wave in terms of probability. Also understand the uncertainty principle relates position and momentum measurements.

Important Concepts

☐ photon
☐ energy of a photon
☐ momentum of a photon
☐ photoelectric effect
☐ work function
☐ Compton effect
☐ de Broglie wavelength

☐ matter wave
☐ Schrödinger's equation
☐ wave function
☐ probability density
☐ uncertainty principle
☐ barrier tunneling

39-1   A New Direction

☐   The remainder of this textbook deals with the physics of the very small: atoms, atomic nuclei, and fundamental particles. These objects do not behave as classical physics might lead us to believe. You must learn some quantum mechanics to understand the physics of the very small.

39–2 Light Waves and Photons

☐ According to quantum theory, the energy in an electromagnetic beam is concentrated in discrete bundles called **photons**. The energy of a single photon associated with a wave of frequency f is $E = hf$, where h is the Planck constant (6.63×10^{-34} J · s). The rate with which energy is carried by a beam is Rhf, where R is the number of photons per unit time that cross a plane perpendicular to their direction of travel. The intensity of a uniform beam is given by $I = Rhf/A$, where A is the cross-sectional area.

39–3 The Photoelectric Effect

☐ When monochromatic light is absorbed, energy is transferred to electrons of the material. Some electrons overcome the potential energy barrier that normally keeps them inside the sample and they leave with various values of kinetic energy. Data taken in a **photoelectric effect** experiment is used to calculate the kinetic energy of those electrons with the greatest kinetic energy.

☐ The experiment shows that, if the frequency is unchanged, the number of electrons ejected depends on the incident intensity but their energies do not. The energy received by each electron does depend on the frequency of the absorbed light.

☐ Classical wave theory cannot explain these results. Averaged over a cycle, the energy carried by a plane wave is spread uniformly over the wave. If the intensity is increased without changing the frequency, each electron should receive more energy and the kinetic energy of the most energetic electron should increase. Electrons are small and require time to collect enough energy from a wave to overcome the potential energy barrier. Experiments show that, even at low intensity, electrons are ejected immediately after the light is turned on.

☐ The photon hypothesis does explain the experimental results. Some fraction of the incident photons interact with electrons and *each* of those that do transfer energy hf to an electron and disappear. If the intensity is increased without increasing the frequency, then the number of interactions and the number of ejected electrons increase. Since each electron that absorbs a photon receives energy hf, the energy of the most energetic electrons does not change. If the frequency increases, then the energy of each photon is greater and each ejected electron receives more energy than before.

☐ The fundamental event here is the interaction of a single quantum of light with a single particle of matter. Suppose a single photon interacts with one of the most energetic electrons in the sample. If ϕ is the energy needed to remove one of these electrons from the sample, then conservation of energy leads to $hf = \phi + K_m$, where K_m is the kinetic energy of the electron after leaving the sample. ϕ is called the **work function** of the sample.

☐ To measure the maximum kinetic energy K_m, the sample is enclosed in a metal collector cup, which is held at a negative electric potential relative to the sample. The potential on the cup is adjusted so the most energetic electrons just fail to reach the cup. If this potential, called the **stopping potential**, is denoted by V_0, then $K_m = eV_0$ gives the maximum kinetic energy.

39–4   Photons Have Momentum

☐ Photons also carry momentum; the momentum of a single photon is $p = h/\lambda$, where λ is the wavelength of the wave. The energy and momentum are related by $E = pc$, where c is the speed of light. The **Compton effect** experiment demonstrates photon momentum.

☐ In this experiment, short-wavelength electromagnetic radiation is scattered by electrons and the wavelength of the scattered radiation is measured. It is found to be longer than the wavelength of the incident radiation and the difference is dependent on the scattering angle. Energy and momentum are transferred from the radiation to the electron.

☐ If the classical wave hypothesis were valid, the wavelength of the scattered radiation would be the same as that of the incident radiation. Electrons oscillate at the frequency of the incident wave and re-radiate at the same frequency.

☐ According to the photon hypothesis, the fundamental interaction, between a single photon and a single electron, can be treated as a collision in which energy and momentum are both conserved. An electron essentially at rest is knocked away from its initial position by a photon. Since the energy of the electron increases in the interaction, the energy of the photon must decrease. Since the photon energy is hf, the frequency of the electromagnetic wave must decrease, and since $\lambda = c/f$, the wavelength must increase.

☐ The diagrams below show the situation before and after the collision. Before the collision, the electron is at rest and a photon associated with a wave of wavelength λ moves toward it along the x axis. After the collision, a photon associated with a wave of wavelength λ' moves off at the angle ϕ and the electron moves off with speed v at the angle θ, as shown.

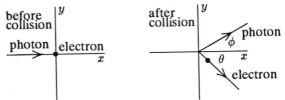

Conservation of the x component of momentum requires that

$$\frac{h}{\lambda} = \frac{h}{\lambda'} \cos\phi + \frac{mv\cos\theta}{\sqrt{1 - (v/c)^2}},$$

conservation of the y component of momentum requires

that

$$0 = \frac{h}{\lambda'} \sin \phi - \frac{mv \sin \theta}{\sqrt{1 - (v/c)^2}} \, ,$$

and conservation of energy requires that

$$\frac{ch}{\lambda} = \frac{ch}{\lambda'} + mc^2 \left(\frac{1}{\sqrt{1 - (v/c)^2}} - 1 \right) \, .$$

Relativistic expression must be used for the electron momentum and kinetic energy. These equations can be solved for the change in wavelength ($\Delta\lambda = \lambda' - \lambda$):

$$\Delta\lambda = \frac{h}{mc}(1 - \cos \phi) \, .$$

☐ In the usual experiment, many photons with the same frequency are incident on many electrons and photons leave at all angles. A detector is moved around a circle centered at the sample and the wavelength of the radiation is measured for each angular position of the detector.

☐ The maximum change in wavelength occurs for a scattering angle of $\phi = 180°$. It is $\Delta\lambda = 2h/mc = 4.86$ pm.

☐ The change in wavelength is independent of the wavelength itself. The Compton effect is important for gamma rays and short-wavelength x rays but not for visible light or microwaves. For long-wavelength radiation, the wavelength shift is an extremely small fraction of the wavelength.

☐ For scattering from bound electrons and nuclei, the change in wavelength is much smaller than for scattering from electrons because the mass of an atom is much greater than that of an electron.

39–5 Light as a Probability Wave

☐ You should understand the quantum mechanical interpretation of an electromagnetic wave: the square of the electric field vector at a point is proportional to the probability per unit time that a photon will be in a small volume centered at that point. Carefully note that physics cannot predict where any photon will be at any time; it can only give the probability a photon will be in some specified region.

☐ Probability waves can interfere destructively and thereby reduce the chance that a photon will get to a given place; they can interfere constructively and thereby increase the chance that a photon will get to a given place.

☐ The equations $E = hf$ and $p = h/\lambda$ each relate a particle property (energy or momentum) and a wave property (frequency or wavelength).

39–6 Electrons and Matter Waves

☐ Particles exhibit interference phenomena. A uniform beam of electrons is reflected from a crystal only if the angle of incidence has certain values. The results are just like those obtained for x rays. Clearly waves are associated with particles.

☐ If a matter wave is sinusoidal, the wavelength λ of the wave is related to the momentum p of the particle by $\lambda = h/p$ and the frequency f of the wave is related to the energy E of the particle by $f = E/h$, where h is the Planck constant. The wavelength of a matter wave is called its **de Broglie wavelength**.

☐ For macroscopic objects, the momentum is usually so large and the wavelength so small that interference and diffraction effects cannot be detected. They are evident, however, for atomic particles.

39–7 Schrödinger's Equation

☐ Quantum physics associates a **wave function** with a particle. This wave function is related to a particle in much the same way as an electromagnetic wave is related to a photon. If the wave function of a particle moving on the x axis is represented by $\Psi(x,t)$, then $|\Psi|^2\,\mathrm{d}x$ gives the probability that the particle is in the small region from x to $x + \mathrm{d}x$. The square of the magnitude of a wave function is called a **probability density**. For one dimensional motion, it is a probability per unit length.

☐ Many matter wave functions are complex. They are the sum of real and imaginary parts. $|\Psi|^2$ is the product of the function with its complex conjugate. If a particle has an energy E, then its wave function can be written as the product of a coordinate-dependent function $\psi(x)$ and a time-dependent function $e^{i\omega t}$, where ω is the angular frequency and is related to the energy by $\omega = 2\pi E/h$. Ψ and π lead to the same probability density: $|\Psi|^2 = |\psi|^2$.

☐ A fundamental equation, called **Schrödinger's equation**, allows us to solve for ψ if the forces acting on the particle are known. It is

$$\frac{\mathrm{d}^2\psi}{\mathrm{d}x^2} = \frac{8\pi^2 m}{h^2}\left[E - E_{\mathrm{pot}}(x)\right]\psi = 0,$$

where E is the energy of the particle, m is its mass, and E_{pot} is the potential energy. For any given situation, if the function E_{pot} is known, this equation is to be solved for the wave function.

☐ If the particle is free, then $E_{\mathrm{pot}} = 0$ and Schrödinger's equation becomes

$$\frac{\mathrm{d}^2\psi}{\mathrm{d}x^2} + \frac{8\pi^2 m}{h^2}E\psi = 0.$$

The wave function for a free particle traveling in the positive x direction is $\psi(x) = Ae^{ikx}$ and the wave function for a free particle traveling in the negative x direction is $\psi(x) = Ae^{-ikx}$, where $k = \sqrt{2mE/\hbar^2}$. These functions are sinusoidal and each has a wavelength λ that is related to k by $k = 2\pi/\lambda$.

39-8 Heisenberg's Uncertainty Principle

☐ If the wave function of a particle does not have the form Ae^{ikx}, then the particle does not have a definite momentum. If we measure the momentum many times, with the particle always in the same state, we obtain a distribution of values. Quantum mechanics can predict the probability that the particle's momentum will be in any given range but it cannot predict the result of any of the momentum measurements. The situation is quite similar to the measurement of position: quantum mechanics can predict the probability that a particle is in a given region of space but it cannot predict its position at any given time.

☐ The position of a particle is localized in space if its wave function is large only over a small region and is zero or nearly zero everywhere else. If the wave function of a particle is changed so the particle's position is more localized, then its position is known with a higher certainty. The particle's momentum then becomes less certain.

☐ In general, if Δx is the uncertainty in the position of a particle and Δp is the uncertainty in its momentum, then $\Delta x \cdot \Delta p \geq \hbar$, where $\hbar = h/2\pi$. Similar relationships hold for each coordinate and the corresponding momentum component.

☐ The Planck constant h appears in the uncertainty relations. If the classical limit is obtained by setting $\hbar = 0$, then the relationship for the x component becomes $\Delta p \cdot \Delta x = 0$. Both the position and momentum can be entirely certain.

39–9   Barrier Tunneling

☐   A matter wave extends into the region beyond a potential energy barrier if the barrier has finite height and width. This means a particle may escape from such a region. Suppose a particle going to the right approaches a potential energy barrier with a height E_{pot} that is greater than its energy E. Classically, the particle cannot travel to the other side of the barrier. Quantum mechanically, it has a non-zero probability of being on either side.

☐   Tunneling through a barrier is described quantitatively by a **transmission coefficient** T and a **reflection coefficient** R. The transmission coefficient is defined so that it gives the probability the particle tunnels through the barrier, while the reflection coefficient gives the probability the particle does not. The sum of the two is one.

☐   If the transmission coefficient is small, it is given by

$$T = e^{-2kL},$$

where

$$k = \sqrt{\frac{8\pi^2 m (E_{pot} - E)}{h^2}},$$

E is the energy of the particle, and L is the width of the barrier. If the barrier is made wider (L is increased), then T decreases and if the barrier is made higher (E_{pot} is increased), then k increases and T decreases. The parameter k depends on $E_{pot} - E$. If the energy of the particle is increased (but is still not greater than E_{pot}), then k decreases and T increases.

NOTES:

Chapter 39:   Photons and Matter Waves

Chapter 40
MORE ABOUT MATTER WAVES

In this chapter you learn some details about matter waves that will prove useful when you study atoms. Learn that the energy of a particle is quantized when the particle is confined in space, as electrons in atoms are. Also learn the mathematical forms of matter waves for an electron in a one-dimensional trap and in a hydrogen atom. Pay attention to the quantum numbers used to designate hydrogen atom states.

Important Concepts

□ trapped electron
□ energy quantization
□ normalized wave function
□ zero-point energy
□ ground state
□ excited state
□ radiative transitions
□ hydrogen atom energies

□ hydrogen atom wave functions
□ Bohr radius
□ radial probability function
□ principal quantum number
□ orbital quantum number
□ orbital magnetic
  quantum number

40–1 Atom Building

□ Quantum mechanics is essential for our understanding of atoms and smaller structures. We must take into account matter waves and their probabilistic interpretation. These can be obtained by solving Schrödinger's equation.

40–2 Waves on Strings and Matter Waves

□ If a particle is confined to a limited region of space, a finite segment of the x axis, say, then its energy is quantized. That is, the energy can have any of a set of definite discrete values. This result is closely related to a similar result for a wave on a string that is clamped at both ends. In the absence of an external driving force, the string can vibrate with any of a set of discrete frequencies, the normal modes or harmonics.

40–3 Trapping an Electron

□ Suppose an electron is trapped on the x axis by infinite potential energy barriers at $x = 0$ and $x = L$. No force acts on it when it is in the interior but when it reaches either end of the trap, an infinite force pushes it toward the interior. Inside the trap ($0 \leq x \leq L$) the coordinate-dependent part of the wave function ψ satisfies the Schrödinger equation

$$\frac{\mathrm{d}^2\psi}{\mathrm{d}x^2} + \frac{8\pi^2 m}{h^2} E\psi = 0 \,.$$

Outside the trap, $\psi = 0$. ψ obeys the boundary conditions $\psi(0) = 0$ and $\psi(L) = 0$.

□ There are many possible wave functions for a particle in a trap. They are quite similar to the functions that describe the displacement of a vibrating string that is fixed at both ends. Inside the trap (from $x = 0$ to $x = L$),

$$\psi_n = \sqrt{\frac{2}{L}} \sin\left(\frac{n\pi x}{L}\right) \,,$$

where n is a positive integer. Outside the trap, $\psi_n = 0$. The integer n distinguishes one function from another and is used to label the functions. It is called a *quantum number*. Notice that $\psi_n = 0$ at $x = 0$ and at $x = L$.

□ If the wave function is ψ_n, then the probability density is

$$|\psi_n|^2 = \left(\frac{2}{L}\right) \sin^2\left(\frac{n\pi x}{L}\right) \,.$$

Fig. 40–4 shows graphs of the probability densities for four states.

☐ The functions have been **normalized**. That is, the constant in front ($\sqrt{2/L}$) is chosen so that

$$\int_{-\infty}^{+\infty} |\psi|^2 \, dx = 1 \,.$$

This condition is derived from the statement that $|\psi|^2 \, dx$ is the probability that the particle is between x and $cx + dx$. Since the particle must be somewhere on the x axis, the sum of probabilities for all segments of the axis must be 1.

☐ An energy is associated with each wave function. If the particle is in the state with wave function ψ_n, then its energy is given by

$$E_n = \frac{n^2 h^2}{8mL^2} \,.$$

The energy is said to be quantized. It can have any of the values given above but it can have no others. This is quite generally true of any particle that is confined to a limited region of space.

☐ Notice that the particle has non-zero energy when it is in its lowest (or ground) state. This energy, which is given by $h^2/8mL^2$, is called its **zero-point energy**. The particle cannot be at rest under any conditions.

☐ Notice that the wave functions are quite similar to the functions that describe the displacement of a vibrating string fixed at both ends. Both the matter wave and the string displacement are the sum of sinusoidal waves with the same frequency and wavelength, traveling in opposite directions. For the standing wave to have zero amplitude at $x = 0$ and $x = L$, the wavelength λ and the width of the trap must be related by $\lambda = 2L/n$. If the traveling matter waves extend along the entire x axis, they would be associated with a free particle that has momentum with magnitude $p = h/\lambda = nh/2L$. The energy is $E = p^2/2m = n^2 h^2/8mL^2$.

□ A trapped particle may absorb a photon and, thus, increase its energy but this happens only if the photon energy equals the difference in energy between the initial state of the particle and some higher state. Suppose the electron is initially in the state with quantum number n_i when it absorbs a photon and goes to the state with quantum number n_f. Then, the photon energy is

$$E_{\text{photon}} = E_f - E_i = \frac{(n_f^2 - n_i^2)h^2}{8mL^2}.$$

Similarly, a particle in an excited state can make the transition to a state with lower energy by emitting a photon. The energy of the photon equals the difference in energy between the initial and final states.

40–4 An Electron in a Finite Well

□ If the barriers that form the walls of the trap are finite, then particle energies are quantized if they are less than the barrier height and are not quantized if they are greater. If $E > E_{\text{pot}}$, where E_{pot} is the barrier height, then the particle is not confined to the trap.

□ The wave functions are sinusoidal inside the trap and tend exponentially to zero in the barriers. Fig. 40–6 shows some probability densities. There is some chance that the particle will be outside the trap, in the region forbidden to it by classical mechanics.

40–5 More Electron Traps

□ Atoms are electron traps. The positive charge of the nucleus confines the electron and causes its energy to be quantized.

□ Crystals with dimensions on the order of nanometers can trap electrons. Such crystals absorb all incident light with wavelength below a certain threshold value that depends on the crystal size and reflect all light with wavelength above the threshold. The frequency of the long wavelength light, and, hence, the energy of its photons, is so small that it cannot excite an electron from a low-lying energy level to a higher level. Thus, the crystal size controls the color of the reflected light.

☐ Quantum dots are constructed by sandwiching a tiny piece of semiconductor between insulating materials. The semiconductor becomes an electron trap. Such devices are used in the electronic and optical communications industry.

40–6 The Hydrogen Atom

☐ The set of wave functions appropriate to an electron in a hydrogen atom differs from the set appropriate to an electron in another situation (trapped in a well, for example) because the force acting on the electron is different for the two cases. The important quantity for a quantum mechanical calculation of a particle wave function is the potential energy function and this is different if different forces act. For hydrogen, which consists of an electron and proton, this function is

$$E_{\text{pot}} = -\frac{e^2}{4\pi\epsilon_0 r} \, ,$$

where r is the separation of the two particles.

☐ The electron is trapped by the proton and quantum mechanics predicts a discrete set of energy levels. They are given by

$$E_n = -\frac{me^4}{8\epsilon_0^2 h^2 n^2} = -\frac{13.6\,\text{eV}}{n^2} \, ,$$

where n is a positive integer. Notice that even in the lowest energy state ($n = 1$), the electron has energy and is not at rest.

☐ Wave functions for electrons in atoms are identified by a set of three quantum numbers: the principal quantum number

n, the orbital quantum number ℓ, and the orbital magnetic quantum number m_ℓ. The first is closely associated with the energy, the second with the magnitude of the orbital angular momentum, and the third with one component of the orbital angular momentum. The values of ℓ are restricted by the value of n. For a given value of n, ℓ can have any integer value from 0 to $n - 1$. The values of m_ℓ are restricted by the value for ℓ. For a given value of ℓ, m_ℓ can have any integer value from $-\ell$ to $+\ell$. For the ground state of hydrogen, $n = 1$, $\ell = 0$, and $m_\ell = 0$. The allowed values of the energy for a hydrogen atom depend only on n, not the other quantum numbers. For other atoms, they also depend on ℓ.

☐ All states with the same value of n (but different values of ℓ and m_ℓ) belong to the same *shell*. All states with the same value of n and the same value of ℓ (but not the same value of m_ℓ) belong to the same *subshell*.

☐ The ground state ($n = 1$) probability density for the electron in a hydrogen atom is

$$|\psi|^2(r) = \frac{1}{\pi a^3} e^{-2r/a} ,$$

where a is the **Bohr radius** (5.292×10^{-11} m). This is the orbit radius predicted by the Bohr model.

☐ If dV is an infinitesimal volume a distance r from the proton, then the probability the electron is in that volume is given by $|\psi|^2 \, dV$. The **radial probability density** $P(r)$ is defined so that $P(r) \, dr$ gives the probability of finding the electron in a spherical shell of width dr a distance r from the proton. Since the volume of the shell is $dV = 4\pi r^2 \, dr$, the radial probability density is $P(r) = 4\pi r^2 |\psi|^2$. The radial probability density for the ground state wave function of hydrogen is

$$P(r) = \frac{4}{a^3} r^2 e^{-2r/a} .$$

Fig. 40–13 shows a graph of this function. The maximum occurs at $r = a$. The fraction of the time the electron is inside a sphere of radius a is about 0.32; the fraction of the time it is outside the sphere is about 0.68.

□ The $n = 1$ state is spherically symmetric; the wave function depends only on the distance r from the proton and not on any angular variables. The same is true of the $n = 2$, $\ell = 0$ state and, in fact, of any state for which $\ell = 0$. Wave functions for the three $n = 2$, $\ell = 1$ states, however, depend on the angular coordinates θ and ϕ as well as on r. But the sum of the probability densities for these three states is spherically symmetric. Thus, the probability distribution for a completely filled subshell is spherically symmetric.

□ A hydrogen atom in an excited state can emit a photon and make a transition to a lower energy state. If n_i is the principal quantum number associated with the initial state and n_f is the principal quantum number associated with the final state, then the energy of the photon is given by

$$E_{\text{photon}} = \left(\frac{me^4}{8\epsilon_0^2 h^2} \right) \left(\frac{1}{n_i^2} - \frac{1}{n_f^2} \right)$$

and the frequency of the associated wave is given by $f = E_{\text{photon}}/h$.

□ Possible transitions are grouped into series, with all members of a series having the same value of n_f. If $n_f = 1$, the transition is a member of the Lyman series; if $n_f = 2$, the transition is a member of the Balmer series; and if $n_f = 3$, the transition is a member of the Paschen series. The series limit of any series is obtained by setting $n_i = \infty$. For any series, this radiation has the greatest frequency and the shortest wavelength of any radiation in the series.

40–7 Quantum Weirdness: An Example

☐ The text describes what is called the Einstein, Podolsky, Rosen thought experiment. An atom sends out two photons in opposite directions, each having a property called X. When X is measured, two values are possible, X_1 and X_2. The system is such that if photon A has the value X_1, then photon B will have the property X_2 and vice versa. We may think of the system as having two possible states, which can be denoted by (AX_1, BX_2) and (AX_2, BX_1). The actual wave function is a combination of these two states, with equal probability of either one occurring. When the photons are far apart, X is measured for photon A. Either result can be obtained. If X_1 is obtained, then the wave function collapses to (AX_1, BX_2). If X is now measured for B, the result is X_2, without exception. The question is: How did photon B know that the state changed to (AX_1, BX_2)?

NOTES:

Chapter 41
ALL ABOUT ATOMS

Here you learn about the structure of atoms and, in particular, about the states of multi-electron atoms. Atomic structure is reflected in the periodic table of chemistry and is fundamental to our understanding of the properties of materials. An important new idea is the Pauli exclusion principle, which greatly influences the structures of atoms and their chemical interactions and determines many important details of the periodic table. The chapter closes with an explanation of how a laser works.

Important Concepts

- ☐ spin angular momentum
- ☐ spin quantum number
- ☐ spin magnetic quantum number
- ☐ magnetic dipole moment
- ☐ Pauli exclusion principle
- ☐ subshell
- ☐ periodic table

- ☐ continuous x-ray spectrum
- ☐ cut-off wavelength
- ☐ characteristic x-ray spectrum
- ☐ Moseley plot
- ☐ laser
- ☐ stimulated emission
- ☐ metastable state
- ☐ population inversion

41–2   Some Properties of Atoms

☐ Some of the properties of atoms that quantum physics is used to explain are:

1. Atomic properties are periodic. The periodic table of chemistry is arranged in rows, with 2, 8, 18, 18, and 32 chemical elements in successive rows. Atomic properties change in a regular way along a row and are quite similar for atoms in the same column. See Fig. 41–2 for a plot of ionization energies as a function of atomic number.

2. Atoms emit and absorb electromagnetic radiation. An atom has a set of discrete allowable energy values and a photon is emitted when an atom goes from a higher energy state to a lower energy state. When a photon is absorbed, the atom goes from a lower state to a higher state. Conservation of energy yields $hf = |E_i - E_f|$, where f is the frequency of the electromagnetic wave, E_i is the initial energy of the atom, and E_f is the final energy.

3. Atoms have angular momentum and magnetic dipole moments. Angular momentum is associated with the orbital motion of an electron and with its spin. Atomic nuclei also have spin angular momentum. A magnetic dipole moment is associated with each of these angular momenta.

41–3 Electron Spin

☐ Electrons (and other particles) have intrinsic angular momenta, which have nothing to do with their motions. Although there is no evidence that an electron is actually spinning, its intrinsic angular momentum is called its **spin angular momentum** or just its spin.

41–4 Angular Momenta and Magnetic Dipole Moments

☐ The magnitude of an electron's orbital angular momentum is quantized. The allowed values are given by

$$L = \sqrt{\ell(\ell + 1)}\,\hbar\,,$$

where ℓ is zero or a positive integer and is called the **orbital quantum number**. The constant $\hbar$ is the Planck constant divided by 2π. When the atom is in a state with principal quantum number n, the quantum number ℓ can take on the values $0, 1, 2, \ldots, n - 1$.

☐ Each component of the angular momentum is also quantized. The z component, for example, may have only the values

$$L_z = m_\ell \hbar\,,$$

where m_ℓ is an integer (positive, negative, or zero) and is called the **magnetic orbital quantum number**. If the orbital quantum number is ℓ, the possible values of m_ℓ run from $-\ell$ to $+\ell$ and include all integer values between.

☐ Since the z component of the angular momentum is quantized, the angle between the angular momentum vector and the z axis can have only certain values, given by $\cos\theta = L_z/L = m_\ell/\sqrt{\ell(\ell+1)}$. For a given value of ℓ, the smallest angle occurs when $m_\ell = \ell$.

☐ A magnetic dipole moment is associated with the orbital motion of an electron. At the atomic level, magnetic dipole moments are conveniently measured in units of the Bohr magneton μ_B ($= eh/4\pi m = 9.274 \times 10^{-24}$ J/T $= 5.788 \times 10^{-5}$ eV/T).

☐ If an electron is in a state with magnetic quantum number m_ℓ, then in terms of m_ℓ and μ_B, the z component of its orbital magnetic dipole moment is $\mu_{\mathrm{orb},z} = -m_\ell\mu_B$. The direction of the dipole moment is opposite to the direction of the angular momentum because the electron is negatively charged.

☐ The magnitude of the spin angular momentum is given by $S = \sqrt{s(s+1)}\hbar$, where s is called the **spin quantum number** and has the value $s = 1/2$ for electrons. The value of the z component is given by $S_z = m_s\hbar$, where the **spin magnetic quantum number** m_s is either $-1/2$ or $+1/2$.

☐ A magnetic dipole moment is associated with spin angular momentum. If an electron has spin magnetic quantum number m_s then the z component of its spin dipole moment is $\mu_{s,z} = -2m_s\mu_B$. Except for a factor 2, this is the same as the relationship between the z components of the orbital dipole moment and the orbital angular momentum.

☐ Spin angular momentum and its quantum number are not predicted by Schrödinger's equation but they are predicted by relativistic quantum theory.

- [ ] The total angular momentum of an atom is denoted by **J** and is the vector sum of the orbital angular momenta and spin angular momenta of all its electrons. Similarly, the magnetic dipole moment of an atom is the vector sum of the orbital and spin dipole moments of all the electrons. The dipole moment may not be parallel to **J**.

41–5 The Stern-Gerlach Experiment

- [ ] This experiment provides an important experimental verification of space quantization. A beam of atoms passes through a non-uniform magnetic field to a screen. If the field is in the z direction and varies with z, then it exerts a force on the atoms in either the positive or negative z direction, depending on whether μ_z is positive or negative. If the angle between the dipole moment μ and the magnetic field **B** is θ, then the force is given by $F_z = \mu(dB/dz)\cos\theta$.

- [ ] Atoms with different orientations of their magnetic moments are deflected in different directions. The number of directions can be counted and used to calculate ℓ.

41–6 Magnetic Resonance

- [ ] **Nuclear magnetic resonance** measurements detect changes in the spin directions of protons in materials when the protons in a constant magnetic field **B** are subjected to a sinusoidal electromagnetic field. If the frequency f and total magnetic field B are related by $hf = 2\mu B$, the electromagnetic wave causes the proton spins to flip and energy is preferentially absorbed from the wave. Here μ is the proton magnetic dipole moment. Resonance data is used in analytical chemistry to identify compounds and in medicine to make images of the interior of a body.

41–7 Building the Periodic Table

- [ ] The same quantum numbers n, ℓ, m_ℓ, and m_s can be used to label electron states in all atoms, even though the corresponding wave functions and energies are different for electrons in different atoms.

☐ Electrons obey the **Pauli exclusion principle**, which states that only a single electron can be assigned to any quantum state. For an atom, at least one of the quantum numbers n, ℓ, m_ℓ, and m_s must be different for each electron.

☐ When an atom is in its ground state, the electrons fill the various states in such a way that the total energy is the least possible. You may start with an atom with atomic number $Z - 1$ and construct an atom with atomic number Z by adding a single proton to the nucleus and a single electron outside. If the atom is in its ground state, the electron goes into the state that gives the atom the lowest possible energy without violating the Pauli exclusion principle.

☐ All electrons with the same values of n and ℓ are said to belong to the same **subshell**. The various values of ℓ are designated by lower case letters when labeling a subshell: $\ell = 0$ is designated s, $\ell = 1$ is designated p, $\ell = 2$ is designated d, $\ell = 3$ is designated f, $\ell = 4$ is designated f, and $\ell = 5$ is designated h. A particular subshell is designated by a symbol such as 3d, which indicates that $n = 3$ and $\ell = 2$. The number of electrons in the subshell is written as a superscript. Thus, $3d^2$ indicates that there are two electrons in the 3d subshell.

☐ The number of states in a subshell is $2(2\ell + 1)$. The orbital and spin angular momentum vectors point in all possible directions for electrons in a filled subshell, making the total orbital angular momentum and the total spin angular momentum both zero. The contribution of a filled orbital to the magnetic dipole moment of the atom is zero.

☐ A neon atom has ten electrons. Two of these occupy the $n = 1$, $\ell = 0$ subshell, two occupy the $n = 2$, $\ell = 0$ subshell, and six occupy the $n = 2$, $\ell = 1$ subshell. The ten electrons completely fill three subshells, so a neon atom in its ground state has a total angular momentum of zero and a magnetic moment of zero. Electrons in filled subshells are strongly bound and interact only weakly with neighboring atoms. Neon is chemically inert.

☐ A sodium atom has eleven electrons. In the ground state, ten of them are in the same subshells as the ten electrons of a neon atom. The other is in the $n = 3$, $\ell = 0$ subshell. The orbital angular momentum and magnetic dipole moment of a sodium atom are due entirely to the spin of this electron. A single electron outside completely filled subshells is weakly bound and readily interacts with neighboring atoms. Sodium is chemically active.

☐ A chlorine atom has seventeen electrons. In the ground state, ten have the same configuration as neon, two fill the 3s subshell and five are in the 3p shell. Chlorine is very active chemically. It combines easily with atoms that have an additional electron outside a closed shell. This electron can occupy the sixth state in the 3p subshell.

☐ An iron atom has twenty-six electrons. Eighteen of them occupy closed subshells. These are the 1s, 2s, 2p, 3s, and 3p subshells. Six are in the 3d subshell and two are in the 4s subshell. Note that the 4s subshell starts filling before the 3d subshell is full. The $3d^6 4s^2$ configuration has a lower energy than the $3d^8$ configuration.

☐ The periodic table of the chemical elements groups atoms with similar chemical properties in the same column, with atoms in neighboring columns of the same row differing by one in the number of electrons. Application of quantum mechanics and the Pauli exclusion principle explains the similarities of the properties of atoms in the same column and the variation of properties from column to column.

41–8 X Rays and the Numbering of the Elements

☐ When highly energetic electrons strike nearly any target, x-ray radiation is produced. The radiation spectrum has two parts: a broad, low-level, **continuous spectrum** and a series of sharp peaks. The sharp peaks can be used to investigate electrons in deep-lying energy states of the atoms and to assign atomic numbers to the chemical elements.

☐ Energetic electrons are produced by accelerating electrons from a hot filament through a potential difference. If the potential difference is V, the kinetic energy of an electron is $K = eV$.

☐ An important feature of the continuous spectrum is the **cut-off wavelength**; for any fixed incident electron energy, no x rays with wavelengths less than λ_{min} are emitted.

☐ After an electron enters the target it may lose some or all of its kinetic energy in a decelerating encounter with a nucleus. An x-ray photon is emitted in one of these events. Radiation with the greatest possible frequency (and shortest possible wavelength) is emitted by those electrons that lose *all* their kinetic energy in a single event. The maximum frequency is, therefore, given by $hf_{max} = eV$, so the shortest possible wavelength is given by $\lambda_{min} = hc/eV$. The value of λ_{min} is independent of the target material.

☐ Radiation of the characteristic spectrum (the sharp peaks) is emitted when an incoming electron knocks another electron from a low-energy state, an $n = 1$ state, for example. An electron from a state with higher energy makes the transition to the empty state and a photon is emitted. For most atoms, the difference in energy of the two states is sufficient to produce a photon in the x-ray region of the electromagnetic spectrum. The K_α x-ray line is produced when electrons fall from an $n = 2$ state (labeled an L state) to an $n = 1$ state (labeled a K state). The K_β line is produced when electrons fall from an $n = 3$ state (labeled an M state) to a K state. Radiation with other frequencies is produced when electrons with still higher energies fall into vacated states.

☐ X-ray emissions can be used to determine the position of a chemical element in the periodic table. Moseley showed that the frequency f of the K_α line, for example, changes in a regular way from element to element in the table: a plot of $\sqrt{f}$ as a function of position in the table is a nearly straight line. Moseley concluded that the order of atoms in the table depends on the number of protons in the nucleus.

☐ You should be able to use the results of quantum theory to understand this result. If there are Z protons in the nucleus, then the effective charge that acts on an electron in a K state is closely approximated by $(Z - 1)e$. According to quantum theory, the energy associated with a deep lying state with principal quantum number n is

$$E_n = -\frac{m(Z-1)^2 e^4}{8\epsilon_0^2 h^2 n^2}.$$

This is the same as the expression for hydrogen atom energies except that e^4 in the numerator has been replaced by $(Z - 1)^2 e^4$. The frequency associated with K_α-photons is given by $(10.2\,\text{eV})(Z - 1)^2$. This shows that $\sqrt{f}$ is proportional to $Z - 1$. A Moseley plot is essentially a graph of $\sqrt{f}$ vs. Z and is nearly a straight line. Z is called the atomic number of the chemical element. The expression given above for E_n gives a very poor approximation to the energies of outer electrons in many-electron atoms but it is quite good for the innermost electrons responsible for x-ray emission.

41–9 Lasers and Laser Light

☐ Laser light has the following important properties:

1. It is highly monochromatic.
2. It is highly coherent.
3. It is highly directional.
4. It can be sharply focused.

41–10 How Lasers Work

☐ Laser light is produced in the following process. An electron in a state with high energy is stimulated by an incoming photon to drop to a state with lower energy and emit a photon. The energy of the incoming photon must match the difference in energy of the two states ($hf = E_2 - E_1$) and the energy of the stimulated photon is exactly the same. After the emission, there are two photons, identical in every way. If there are sufficient electrons in the higher energy state, the process continues and the number of identical photons quickly multiplies.

☐ Laser light is monochromatic because all the photons produced by stimulation have the same energy; all the waves associated with them have the same frequency. Laser light is highly coherent because all the waves associated with the stimulated photons have the same phase. Laser light does not spread significantly because all the stimulated photons travel in the same direction. Actually some spreading does occur because the waves are diffracted as they pass through the window of the laser.

☐ To produce laser light, the upper of the two states must be **metastable**. Atoms must remain in those states until the transition to a lower state is stimulated by a photon. In addition, the number of atoms in the upper states must be greater than the number in the lower states, so there are more downward than upward transitions. Since there are normally more atoms in lower energy states than in higher energy states, the desired condition is called **population inversion**.

☐ In a helium-neon laser, two neon energy levels, both above the ground state, are responsible for laser emission. An electric current causes, through collisions, a helium level near the upper neon level to become occupied. When an excited helium atom collides with a neon atom in its ground state, energy is transferred to the neon atom and its upper level becomes occupied. The lower level, being above the ground state, is essentially unoccupied and so collisions with excited helium atoms bring about population inversion. The excited neon atoms are stimulated to emit photons by photons that are already present.

NOTES:

Chapter 42
CONDUCTION OF ELECTRICITY IN SOLIDS

The ideas of modern physics are used to understand one of the important properties of solids, their electrical conductivity. Pay attention to distinctions between metals, insulators, and semiconductors. These materials differ greatly in the fraction of their electrons that participate in electrical conduction and in the changes that occur in their resistivities when the temperature changes. You should understand how the differences come about. Later sections of the chapter are devoted to solid-state devices, so pervasive in modern technology.

Important Concepts

☐ insulator
☐ metal
☐ semiconductor
☐ energy band
☐ energy gap
☐ Fermi energy
☐ density of states
☐ occupation probability
☐ density of occupied states

☐ conduction band
☐ valence band
☐ hole
☐ doping
☐ p-n junction
☐ junction rectifier
☐ light-emitting diode
☐ solid state laser
MOSFET

42–2 The Electrical Properties of Solids

☐ **Electrical insulators** have very few electrons that are free to contribute to an electrical current; their resistivities are large. **Metals,** with large conduction electron concentrations, are good conductors. They have low resistivities. In between are **semiconductors**. Metals have positive temperature coefficients of resistivity; their resistivities increase with increasing temperature. The temperature coefficient of resistivity of a typical semiconductor is larger than that of a metal and it is negative.

42–3 Energy Levels in a Crystalline Solid

☐ Allowed energies for electrons in a crystalline solid form **bands**, large groups of closely spaced levels separated by gaps. Typically, bands are a few electron volts wide; gaps may range from somewhat less than an electron volt to several electron volts. The number of quantum mechanical states in a band is a small multiple of the number of atoms in the crystal. Energy bands arise as atoms are brought close together and the wave functions originally associated with different atoms overlap; several neighboring atoms exert electrical forces on any electron.

☐ The distribution of electrons among the quantum states determines if a substance is a metal or not.

42–4 Insulators

☐ Electrons obey the Pauli exclusion principle: there is at most one electron in each state. At temperature $T = 0\,\text{K}$, the N electrons of a solid fill the N states with the lowest energy. For an insulator, the number of electrons is exactly right to completely fill an integer number of bands, so there are no partially filled bands.

☐ A completely filled band makes no contribution to an electrical current because every electron can be paired with another that is traveling with the same speed but in the opposite direction. An electric field cannot change this situation because no empty states are nearby for the electron to occupy. The gap between the highest filled band and the empty band above is so large that essentially no electrons are excited across it by thermal agitation or by an applied electric field.

42–5 Metals

☐ For a metal, the states in one band are partially occupied. States in lower energy bands are essentially all occupied and states in higher energy bands are essentially all unoccupied. Electrons in the partially filled band are the conduction electrons and are responsible for the current when an electric field is turned on.

☐ The energy of the highest occupied state at $T = 0\,\mathrm{K}$ is called the **Fermi energy** and is denoted by E_F. For a metal, nearly all the energy of an electron is kinetic energy, so the speed of an electron at the Fermi level is $v_F = \sqrt{2E_F/m}$. This is called the **Fermi speed** and typically has a value of about $10^6\,\mathrm{m/s}$.

☐ In the absence of an electric field in the metal, the average electron velocity is zero because, for every electron traveling in any direction, another is traveling with the same speed in the opposite direction. Because the band is partially filled, there are unoccupied states with energy near the Fermi energy and with velocity opposite the direction of the field. An electric field causes a small fraction of the conduction electrons to occupy these states. A current results.

☐ If the electrons did not suffer collisions with atoms of the solid, their velocities would continue to increase in the direction opposite the field. Electrons with energies near the Fermi energy, however, do suffer collisions. Electrons with less energy do not because there are no nearby empty states for them to occupy.

☐ The resistivity of a metal is determined by the number n of electrons per unit volume in the partially filled band and by the average time τ between collisions (the **relaxation time**): $\rho = m/ne^2\tau$, where m is the mass of an electron and e is the magnitude of its charge.

☐ The **density of states** $N(E)$ is defined so $N(E)\,dE$ gives the number of quantum mechanical states per unit volume of sample that have energies between E and $E + dE$. For a free-electron metal, it is given by

$$N(E) = \frac{8\sqrt{2}\pi m^{3/2}}{h^3} E^{1/2}.$$

☐ Not all states are filled. The probability that at absolute temperature T a state with energy E contains an electron is given by the occupancy probability

$$P(E) = \frac{1}{e^{(E-E_F)/kT} + 1},$$

where k is the Boltzmann constant and T is the absolute temperature. This function takes the Pauli exclusion principle into account: no state may be occupied by more than one electron. The **density of occupied states** N_0, which is the number of electrons per unit volume of sample per unit energy interval, is given by the product of $N(E)$ and $P(E)$: $N_0(E) = N(E)P(E)$.

☐ At the absolute zero of temperature, the probability that a state with energy less than E_F is occupied is one; these states are guaranteed to be filled. The probability that a state with energy greater than E_F is occupied is zero; these states are guaranteed to be empty. The Fermi energy at $T = 0\,\text{K}$ is determined by the condition that the number of occupied states per unit volume in the partially filled band is the same as the number of conduction electrons per unit volume in the metal. The determining condition is written mathematically as

$$n = \int_0^{E_F} n(E)\,\mathrm{d}E,$$

where the energy for the bottom of the band was taken to be zero. Once the integral is evaluated, the result can be solved for the Fermi energy:

$$E_F = \left(\frac{3}{16\sqrt{2}\pi}\right)^{2/3} \frac{h^2}{m} n^{2/3} = \frac{0.121h^2}{m} n^{2/3}.$$

42–6 Semiconductors

☐ There are precisely enough electrons in a semiconductor to completely fill all the states in an integer number of bands, with none left over. At $T = 0\,K$, the most energetic electron is in the highest energy state of one of the bands and is separated in energy from the empty state above it by a gap. The highest filled band is called the **valence band** and the lowest empty band is called the **conduction band**.

☐ Since a completely filled band does not contribute to an electrical current, a semiconductor is an insulator at $T = 0\,K$. Electrons must be promoted across the gap before an electric field will generate a current.

☐ At higher temperatures, electrons are thermally promoted across the gap from the valence to the conduction band. Since the gap for a semiconductor is much less than the gap for an insulator, the conduction electron concentration is much greater for a semiconductor than for an insulator at the same temperature. On the other hand, it is much less for a semiconductor than for a metal. Even at high temperatures, the resistivity of a typical semiconductor is much greater than that of a metal.

☐ When electrons have been excited to the conduction band of a semiconductor, both the conduction and valence bands are partially filled and both contribute to the current when an electric field is turned on. Rather than deal with the contributions of the vast number of electrons in the valence band, the band is thought to consist of a much smaller number of fictitious particles, called **holes**, one for each empty state. These particles have positive charge and are accelerated in the direction of an applied electric field.

☐ The relaxation time decreases somewhat as the temperature increases, but both the number of conduction band electrons and the number of valence band holes increase dramatically with temperature. As a result, the resistivity of a semiconductor decreases as the temperature increases.

42–7 Doped Semiconductors

☐ The number of conduction band electrons or the number of valence band holes can be greatly increased by **doping** a semiconductor: that is, by adding certain impurity atoms.

☐ To increase the number of electrons in its conduction band, a semiconductor is doped with **donor** impurity atoms. Each such atom contributes an electron in a hydrogen atom-like state around the impurity, with energy in the gap between the valence and conduction bands. It is easily promoted to the conduction band by thermal agitation. Semiconductors that are doped with donors are said to be n-type semiconductors.

☐ To increase the number of holes in its valence band, a semiconductor is doped with **acceptor** atoms. Each of these contributes an empty hydrogen atom-like state with energy in the gap. Electrons can easily be promoted from the valence band to impurity states, thereby creating a hole in that band. Semiconductors that are doped with acceptors are said to be p-type semiconductors.

42–8 The p-n Junction

☐ The basic building block of nearly all semiconductor devices is the p-n junction, consisting of p-type and n-type semiconducting materials in contact. Electrons from the n-type material diffuse into the p-type material, where they fall into holes. Holes from the p-type material diffuse into the n-type material, where electrons combine with them. The electron and hole currents, called **diffusion currents**, are in the same direction, from the p side toward the n side.

☐ Electron and hole diffusion leaves a narrow region near the boundary on the n side with positively charged ions, the donors that have lost their electrons, and a narrow region near the boundary on the p side with negatively charged ions, the acceptors that have gained electrons. As a result, an electric field exists near the boundary. It pushes electrons toward the n side and holes toward the p side. This is the **drift current**. If the circuit is not completed, the drift and diffusion currents cancel and the net current is zero. The difference in the electric potential of the two sides is called the **contact potential difference**. The region in which the electric field exists is called the **depletion region** because it has few charge carriers.

42–9   The Junction Rectifier

$\square$ p-n junctions are often used as **rectifiers**. An *ideal* rectifier has infinite resistance for current in one direction and zero resistance for current in the other direction. Rectifiers are used to change an alternating current into a direct current.

$\square$ When a source of emf is connected to a p-n junction, with the positive terminal at the p side, the junction is said to be *forward biased*. The electrical resistance of the junction is small and the current is large. If the positive terminal of the emf is connected to the n side, the junction is said to be *back biased*. The electrical resistance is large and the current is small.

$\square$ A forward bias on a p-n junction lowers the potential barrier and increases the diffusion current dramatically. A back bias on the junction raises the potential barrier and decreases the diffusion current. When the junction is back biased, holes flow from the n to the p side and electrons flow from the p to the n side. Because there are so few holes on the n side and so few electrons on the p side, the current is severely limited.

42–10   The Light-Emitting Diode (LED)

$\square$ If conditions are right, light is emitted from a semiconductor when an electron falls from the conduction band to the valence band. In many cases, the energy lost by the electron equals the energy gap between the two bands. If the gap is E_g, then the frequency of the light is $f = E_g/h$ and its wavelength is $\lambda = hc/E_g$. For most semiconductors, this is infrared or red light. Forward-biased p-n junctions are used to construct light-emitting diodes so that the number of photons emitted is much greater than the number absorbed. The emission of light by recombination of electrons and holes is also the basis of **solid state lasers**.

42–11   The Transistor

☐   Field-effect transistors make use of the dependence of the depletion region width on an applied bias. Two n-type regions, for example, are imbedded in a p-type semiconductor and are connected by a channel of n type material. A potential difference applied to the regions generates a current in the channel. The width of the channel, and hence its resistance, is controlled by a back bias applied to the junction formed by the channel and the substrate. Large changes in the current are produced by small changes in the bias (or gate) potential, so the device can be used for amplification.

NOTES:

Chapter 43
NUCLEAR PHYSICS

A nucleus takes up only an extremely small fraction of the atomic volume but accounts for most of the mass of an atom. You will learn about the constituents of nuclei and about nuclear decay. Pay attention to the energy considerations that tell if a nucleus is stable or not and, if it is not, learn to find the energy of the decay products. Also learn about the mathematics of random decay.

Important Concepts

- ☐ atomic nucleus
- ☐ nuclide
- ☐ isotope
- ☐ atomic number
- ☐ neutron number
- ☐ mass number
- ☐ strong nuclear force

- ☐ disintegration constant
- ☐ binding energy
- ☐ half life
- ☐ disintegration energy
- ☐ alpha decay
- ☐ beta decay
- ☐ nuclear models

43–1 Discovering the Nucleus

☐ Rutherford's experiment consists of firing a beam of alpha particles at a gold foil. Most are deflected through small angles, but a few are deflected by nearly 180°. This is possible only if the nucleus is extremely small and positively charged.

43–2 Some Nuclear Properties

☐ Nuclei consist of **nucleons**, a term that encompasses both protons and neutrons. The **atomic number** (Z) is the number of protons; the **neutron number** (N) is the number of neutrons, and the **mass number** (A) is the total number of nucleons. The word **nuclide** is used to designate a nuclear species. Two nuclides are different if they differ in either their atomic or neutron numbers. Two nuclides that have the same atomic number but differ in neutron number are called **isotopes**.

☐ Nucleons attract each other via the strong nuclear force, an extremely strong force with an extremely short range, about 10^{-15} m. In addition, protons repel each other via electrical forces. As a result, stable nuclei with small mass numbers have the same number of neutrons as protons but stable nuclei with large mass numbers have more neutrons than protons. The extra neutrons participate in the strong nuclear interactions, which hold the nucleus together, but not in the electrical interactions, which tend to tear it apart.

☐ The surface of a nucleus is somewhat ill defined but an average radius R can be measured. It is given by $R = R_0 A^{1/3}$, where A is the mass number and R_0 is about 1.2 fm. A femtometer is 10^{-15} m.

☐ Nuclear masses are commonly measured in **atomic mass units** (abbreviated u): 1 u is about 1.661×10^{-27} kg. The mass of a nucleus in u, rounded to the nearest integer, is its mass number. Because the volume of a nucleus and its mass are both proportional to A, the densities of all nuclei are nearly the same.

☐ The mass of a stable nucleus is less than the sum of the masses of its constituents. The difference accounts for the binding together of the nucleons. If Δm is the mass difference, then the **binding energy** is given by $E = \Delta m\, c^2$, where c is the speed of light. This is the energy that must be given a nucleus to separate it into its constituent particles, well-separated and at rest. Atomic, not nuclear, masses are usually tabulated. The mass of an appropriate number of electrons must be subtracted from the atomic mass to obtain the nuclear mass, or else the mass of an atom with Z protons and N neutrons must be compared to the sum of the masses of Z hydrogen atoms and N neutrons.

☐ Nuclides in the vicinity of iron have the greatest binding energy per nucleon; the binding energy per nucleon drops as A becomes either larger and smaller. See Fig. 43–6. Energy is released in a nuclear fusion event, when two nuclei with low mass numbers combine to form a single nucleus. Energy is also released in a nuclear fission event, when a nucleus with high mass number breaks into smaller fragments.

☐ The internal energy of a nucleus has a discrete set of allowed values, a different set for each nuclide. The difference in energy of adjacent low-lying states is on the order of MeV and when a nucleus changes state from a higher to a lower energy, the photon emitted is in the gamma-ray portion of the electromagnetic spectrum.

☐ Nuclei also have intrinsic angular momenta and magnetic dipole moments. Values of the angular momentum are the same as for electrons in atoms, but nuclear magnetic dipole moments are smaller by a factor of about 1000 because nuclear masses are that much greater than the electron mass.

43–3 Radioactive Decay

☐ An unstable nucleus may spontaneously turn into another nucleus with the emission of one or more particles. Alpha decay involves the emission of a helium nucleus and beta decay involves the emission of either an electron or a positron. The mass of the parent nucleus is greater than the sum of the masses of the decay products.

☐ Although any nucleus in a collection of identical unstable nuclei might decay in any given time interval, which will actually decay cannot be predicted. The number that decay in any small time interval Δt is proportional to the number N of undecayed nuclei present and to the interval itself: $\Delta N = -\lambda N \, \Delta t$, where the constant of proportionality λ is called the **disintegration constant**. The negative sign appears because ΔN is negative (N decreases). In the limit as $\Delta t \to 0$, this equation becomes

$$\frac{dN}{dt} = -\lambda N$$

and its solution is $N(t) = N_0\, e^{-\lambda t}$, where N_0 is the number of undecayed nuclei at time $t = 0$.

☐ The decay rate or activity R also follows an exponential law:

$$R = -\frac{dN}{dt} = R_0\, e^{-\lambda t}\,,$$

where R_0 is the decay rate at $t = 0$. The exponent is the same as the exponent in the law for N. Activities are often measured in becquerels ($1\,\text{Bq} = 1$ disintegration per second.

☐ A radioactive decay is often characterized by its **half-life** τ, the time for half the undecayed nuclei initially present to decay. It is related to the disintegration constant by $\tau = (\ln 2)/\lambda$.

☐ The reduction in mass that occurs with a decay, multiplied by c^2, gives the **disintegration energy**, denoted by Q. This energy appears as the kinetic energy of the decay products and as the excitation energy of the daughter nucleus, if it is left in an excited state.

43–4 Alpha Decay

☐ When a nucleus decays by emission of an alpha particle, the daughter nucleus has an atomic number that is two less than that of parent nucleus, a neutron number that is two less and a mass number that is four less.

☐ The half-lives of alpha emitters range from less than a second to times that are longer than the age of universe. The combination of the strong nuclear force of attraction and the electrical repulsion of the daughter nucleus for the alpha particle leads to a high potential energy barrier that tends to hold the alpha inside. Classically, it can never escape because its energy is less than the barrier height, but quantum mechanical tunneling is possible. The half-life depends sensitively on the height and width of the barrier.

43–5 Beta Decay

☐ In β^- decay, a neutron is converted to a proton, with the emission of an electron and a neutrino. The daughter nucleus has an atomic number that is one greater than that of the parent nucleus, a neutron number that is one less, and a mass number that is the same. In β^+ decay, a proton is converted to a neutron with the emission of a positron and a neutrino. The daughter nucleus has an atomic number that is one less than that of parent nucleus, a neutron number that is one greater, and a mass number that is the same. Neutron-rich nuclides tend to decay by β^- emission while proton-rich nuclides tend to decay by β^+ emission. The β particle and neutrino are created in the decay process; they do not exist before the decay.

☐ The disintegration energy is shared by the β particle, neutrino, and recoiling daughter nucleus, with nearly all of it going to the first two. For identical parent nuclei, the energies of the emitted β particles span a wide range, but the sum of the β and neutrino energies is always the same.

☐ The mass of a neutrino is difficult to measure and may be zero. Neutrinos interact only extremely weakly with matter. Enormous numbers, mostly from the Sun, pass through Earth and our bodies every second.

43–6 Radioactive Dating

☐ Radioactive decay is used to date the deaths of ancient organisms. The atmosphere contains a small amount of radioactive carbon, which enters all living things. When an organism dies, the nuclei lost to decay are not replenished and the time of death can be fixed by measuring the amount of undecayed radioactive carbon left in the organism.

43-7 Measuring Radiation Dosage

☐ The gray (Gy) is used to measure the energy per unit mass of target actually delivered to a target: 1 Gy is equivalent to 1 J/kg delivered. The **sievert** (Sv) is used to measure the dose equivalent and takes into the account the biological effect of the radiation. To find the dose equivalent in sieverts, the dose in grays is multiplied by the *relative biological effectiveness factor*, a factor that is tabulated for various radiations in handbooks.

43-8 Nuclear Models

☐ The **collective model** treats a nucleus as a drop of liquid. It is useful for comparisons of the masses and binding energies of various nuclei and in discussions of nuclear fission. The **independent particle model** treats the nucleons in a nucleus as independent particles. Protons and neutrons have their own sets of energy levels and quantum numbers, like electrons in atoms. The model predicts the existence of shells, which are particularly stable when completely filled. The **combined model** treats closed shells as a liquid drop and nucleons outside as independent particles.

NOTES:

Chapter 44
ENERGY FROM THE NUCLEUS

Both the fission of a heavy nucleus into two lighter nuclei and the fusion of two light nuclei into a heavier nucleus convert internal energy associated with nucleon-nucleon interactions into kinetic energy of the products. In each case, your chief goal should be to understand the basic process. Learn what the products are, what energy is released, and what inhibits the process. In addition, you will learn about occurrences of the phenomena in nature and about human attempts to produce and control nuclear energy in a sustained fashion.

Important Concepts

☐ nuclear fission   ☐ delayed neutron

☐ fission fragment   ☐ nuclear fusion

☐ prompt neutron   ☐ proton-proton cycle

44–2 Nuclear Fission: The Basic Process

☐   In the basic fission process, a thermal neutron is absorbed by a heavy nucleus (^{235}U, say) to form a compound nucleus (^{236}U*) and the compound nucleus breaks into two medium-mass nuclei (fission fragments), with the emission of one or more neutrons. The fission fragments may also emit neutrons and undergo beta decay.

☐   Thermal neutrons, with energies of about 0.04 eV, are used because the chance of capture is much better than for fast neutrons.

☐   Starting with the same nucleus, the mass numbers of the fragments may be different for different fission events and experiments give the relative probabilities for the various possible outcomes. Look at Fig. 44–1 of the text to see the

Chapter 44:   Energy from the Nucleus

distribution of fragments for the fission of ^{236}U. The fragments are not usually identical. The most likely fragments have mass numbers around 95 and 137.

☐ Fission takes place in several steps. First, the compound nucleus breaks into what are called the primary fragments and these fragments immediately emit one or more neutrons, called the **prompt neutrons**. The resulting fragments are usually still neutron-rich and they decay by β^- emission. This decay often leaves a fragment nucleus in an excited state and if the excitation energy is sufficient, it may decay via the emission of more neutrons, although most times the decay is via gamma emission. Neutrons, if they are emitted, are called **delayed neutrons**.

☐ The disintegration energy is given by $Q = \Delta m\,c^2$, where Δm is the change in the total mass. Consider the fission of the heavy compound nucleus F into the fragments X and Y, with the production of b neutrons: F $\rightarrow$ X + Y + bn. Let m_F be the mass of F, m_X be the mass of X, m_Y be the mass of Y, and m_n be the mass of a neutron. Then, the disintegration energy Q is given by $Q = (m_F - m_X - m_Y - bm_n)c^2$. Q is positive, indicating the release of energy. Most appears as the kinetic energy of the fragments but some appears as the kinetic energy of the neutrons, and if X and Y are the final stable nuclei, some appears as the kinetic energies of electrons and neutrinos produced in the β decays.

44–3   A Model for Nuclear Fission

☐ The collective model is used to discuss fission. After the thermal neutron is captured, the fragments are formed and pull apart from each other, much as a single drop splitting into two.

☐ Not all neutron-rich heavy nuclides fission. The absorbed neutron must supply enough energy for the separating fragments to overcome the energy barrier produced by their strong mutual attraction. Once the fragment separation is beyond the peak in the potential energy curve, the potential energy decreases with separation because the fragments are both positively charged and repel each other.

☐ The most energy that can be supplied by a thermal neutron is equal to its binding energy E_n. If this energy is not about the same as the barrier height E_b or greater, fission does not occur and the compound nucleus loses the excitation energy by means of gamma emission. Fission is favored if $E_n \geq E_b$.

44–4   The Nuclear Reactor

☐ To produce a large number of fission events in a reactor, the fissionable material is made to undergo a **chain reaction**: neutrons produced in one fission event are used to initiate other fission events.

☐ If fission takes place in a solid or liquid, most of the released energy appears as thermal energy. In a reactor power generator, the energy is used to raise the temperature of water and the resulting steam is used to drive a turbine.

☐ Neutron production is proportional to the volume of the reactor; the number of neutrons that leak out is proportional to the surface area. Neutron leakage from a reactor is controlled by making the reactor large.

☐ Neutrons produced in fission events are energetic and must be slowed before they can trigger other events. The fuel is mixed with a **moderator** (usually water), so fast neutrons are slowed by collisions with protons in the moderator.

☐ As a neutron slows, it passes a critical energy range in which it is particularly susceptible to nonfissionable capture by a ^{238}U nucleus. The fuel and moderator are clumped so neutrons pass this critical energy range while they are in the moderator and they are slowed to thermal energies before entering the fuel.

☐ If the number of neutrons present in a reactor remains constant with time, the reactor is said to be **critical**; if the number decreases, it is said to be **subcritical**; and if the number increases, it is said to be **supercritical**. The value of the **multiplication factor** k indicates the tendency. If N neutrons that will participate in fission events are present at some time, then after the fission events occur, there will be kN neutrons present to participate in the next round of fission events. The value of k is one if the reactor is critical, less than one if it is subcritical, and greater than one if it is supercritical. For steady power generation, $k = 1$.

☐ The multiplication factor is varied by inserting and withdrawing **control rods**, which readily absorb neutrons. To increase the number of fission events per unit time, some rods are withdrawn for a short time, during which k is greater than one. The rods are then inserted so k becomes one again. Because some neutron emissions are delayed, the rods are effective even though there is a response time for adjusting them.

44–6 Thermonuclear Fusion: The Basic Process

☐ Nuclear fusion occurs when two light nuclei fuse together to form a single heavier nucleus. For low mass numbers, the binding energy per nucleon is much greater for heavier nuclei than for lighter. The heavy nucleus has much less internal energy and the excess appears as the kinetic energy of the products. Products include the final nucleus and perhaps neutrons, β particles, and photons.

☐ The light nuclei are positively charged and repel each other electrically. They must overcome (or tunnel through) a potential energy barrier before they get close enough for the strong force of attraction to be effective. In a hot plasma of protons or other light nuclei, there are many nuclei with sufficient kinetic energy to fuse, even if the mean kinetic energy is much less than the barrier height. Look at Fig. 20–7 to see the distribution of speeds for molecules in a gas.

Chapter 44:   Energy from the Nucleus

44–7 Thermonuclear Fusion in the Sun and Other Stars

☐ The proton-proton cycle of fusion processes are responsible for the internal energy of the Sun and other stars. It proceeds by the following steps:

$^1H + {^1H} \rightarrow {^2H} + e^+ + \nu$ Two protons fuse to form a deuteron. A positron (e^+) and a neutrino (ν) are emitted. The positron annihilates with a free electron, producing two photons.

$^2H + {^1H} \rightarrow {^3He} + \gamma$ A deuteron and a proton fuse to form a light helium nucleus (with only one neutron) and another photon is emitted.

$^3He + {^3He} \rightarrow {^4He} + {^1H} + {^1H}$ Two light helium nuclei fuse to form an alpha particle and two protons.

The overall process can be written $4^1H + 2e^- \rightarrow {^4He} + 2\nu + 6\gamma$.

☐ The disintegration energy for the entire process is about 26.7 MeV. About 0.5 MeV is carried away by the neutrinos, but most is available to maintain the internal energy and keep the Sun shining. It is estimated that the proton-proton cycle can continue for about 5×10^9 y before the Sun's hydrogen is gone.

☐ When all the hydrogen has been fused into helium, the Sun's core will collapse and as gravitational potential energy is converted to kinetic energy, it will get hotter. In the hot dense core, helium can fuse into heavier elements. Chemical elements with mass numbers beyond $A = 56$ cannot be formed in fusion events because the binding energy per nucleon decreases beyond this mass number.

44–8 Controlled Thermonuclear Fusion

☐ Currently, attempts are being made to produce sustained controlled fusion for purposes of generating electric power. The goal is to maintain a high-temperature, high-density gas of particles for sufficient time that a significant number of fusion events take place. If n is the particle concentration and τ is the confinement time, then **Lawson's criterion** for a successful fusion reactor is $n\tau \geq 10^{20}\,\text{s} \cdot \text{m}^{-3}$. In addition, the temperature must be high enough to allow particles to overcome the Coulomb barrier to their fusion.

☐ A tokamak makes use of a magnetic field to maintain high particle concentrations and to keep the hot plasma away from walls. Magnetic confinement methods seek to obtain a long confinement time, although particle concentrations are low.

☐ The inertial confinement method uses a high-power laser beam to ionize and compress small pellets of deuterium and tritium. Fusion takes place in the core of the pellet. This method uses a short confinement time but a high particle concentration in an attempt to reach the critical value of the Lawson number $n\tau$.

NOTES:

Chapter 44:   Energy from the Nucleus

Chapter 45
QUARKS, LEPTONS, AND THE BIG BANG

This chapter deals with two closely related topics: the most fundamental constituents of matter and the evolution of the universe from its start at the Big Bang. Pay attention to the properties of particles and to the mechanisms by which they interact. This information is used to unravel the history of the early universe. Also learn of the evidence for the Big Bang and for dark matter, on which the future of the universe depends.

Important Concepts

- ☐ fermion
- ☐ boson
- ☐ hadron
- ☐ meson
- ☐ baryon
- ☐ lepton
- ☐ antiparticle
- ☐ lepton numbers
- ☐ baryon number
- ☐ strangeness

- ☐ eightfold way
- ☐ quark
- ☐ messenger particle
- ☐ W and Z particles
- ☐ gluon
- ☐ expansion of the universe
- ☐ Hubble's law
- ☐ microwave background radiation
- ☐ dark matter
- ☐ Big Bang

45–2   Particles, Particles, Particles

☐   Intrinsic angular momentum (spin) is used to classify particles. The maximum value of any cartesian component of the spin is $S_z = m_s \hbar$, where m_s is either an integer or half integer and $\hbar$ is the Planck constant divided by 2π. Particles that have m_s equal to 0 or an integer are called **bosons** while particles that have m_s equal to half an odd integer are called **fermions**. Fermions obey the Pauli exclusion principle; bosons do not. Electrons, protons, and neutrons are fermions; photons and pions are bosons.

☐ Particles are also classified according to the types of interactions they have with other particles. Particles that interact via the strong nuclear force are called **hadrons**; particles that do not interact via the strong nuclear force but do interact via the weak nuclear force are called **leptons**. Protons, neutrons, and pions are hadrons; electrons, positrons, and neutrinos are leptons.

☐ Hadrons are further categorized according to their spin. Particles that interact strongly *and* have integer (or zero) spin, such as the pion, are called **mesons**. Particles that interact strongly *and* have half-integer spin, such as the neutron and proton, are called **baryons**.

☐ An **antiparticle** is associated with each particle. A particle and its antiparticle have the same mass and spin but charges of opposite sign (if charged). Other quantum numbers also have different signs for a particle and its antiparticle. Except for the positron (e^+), an antiparticle is denoted by a bar over the symbol for the particle: $\bar{p}$ represents an antiproton, for example.

45–3 An Interlude

☐ Energy, charge, momentum, and angular momentum are conserved in all high energy reactions and decays. Mass is not necessarily conserved.

45–4 The Leptons

☐ Counting the neutrinos but not the antiparticles, there are six members of the lepton family. They are: the electron (e), the muon (μ), the tauon (τ), the electron neutrino (ν_e), the muon neutrino (μ_μ), and the tauon neutrino (μ_τ). All are fermions. The electron, muon, and tauon are charged and participate in electromagnetic interactions; neutrinos are not charged. No evidence of any internal structure has been observed for any lepton. They are thought to be truly fundamental.

☐ Electrons and electron neutrinos are assigned an electron lepton number of +1; positrons and electron antineutrinos are assigned an electron lepton number of −1; all other particles have electron lepton numbers of 0. In a similar manner, muon lepton numbers and taon lepton numbers are also assigned. In every decay or reaction, each of the lepton numbers is conserved: the sum for all particles before an event is the same as the sum for all particles after the event, even if the particles change identities.

45–5 Another Conservation Law

☐ **Baryon number** $B = +1$ is assigned to each baryon, baryon number $B = -1$ is assigned to each antibaryon, and baryon number $B = 0$ is assigned to every other particle. The sum of the baryon numbers before an event is always equal to the sum of the baryon numbers after the event.

☐ No similar conservation law holds for mesons. Mesons can be created or destroyed in any number without violating a conservation law.

45–6 Still Another New Conservation Law

☐ **Strangeness** S is a particle quantum number that is conserved in strong and electromagnetic interactions but not in weak interactions. Particles are assigned strangeness S by observing the results of decays and interactions. K^+ and K^0, for example, have $S = 1$, K^- and $\overline{K}^0$ have $S = -1$.

☐ When a decay or reaction takes place via either the strong or electromagnetic interaction, the total strangeness of the products must be the same as the total strangeness of the original particles. If the decay or reaction takes place via the weak interaction, the total strangeness may change.

45–7 The Eightfold Way

☐ When the strangeness and charge quantum numbers of the eight baryons with spin quantum number $\frac{1}{2}$ are plotted, with the charge on a sloping axis, the result is a hexagonal pattern. The same pattern is formed by the nine mesons with spin zero. See Fig. 45–4. All particles can be arranged in this pattern or a triangular pattern of ten particles. These **eightfold way patterns** are related to the particles in much the same way as the periodic table of chemistry is related to atoms.

45–8   The Quark Model

☐ Baryons and mesons are thought to be made of smaller, more fundamental particles called **quarks**. Every meson is a combination of a quark and an antiquark, whereas every baryon is a combination of three quarks and every antibaryon is a combination of three antiquarks.

☐ The baryon number of a quark is $+1/3$ and that of an antiquark is $-1/3$, so the model predicts that the baryon number of a meson is 0, the baryon number of a baryon is 1, and the baryon number of an antibaryon is -1, in agreement with observation.

☐ Six quarks (and associated antiquarks) have been discovered, in the sense that they are required for the construction of observed mesons or baryons. Many physicists believe no others exist. The lowest mass mesons and baryons consist of u (up), d (down), and s (strange) quarks, along with the associated antiquarks. The quark content of the π^- meson, for example, is $u\bar{d}$; the quark content of the π^+ meson is $d\bar{u}$. Kaons, which are strange mesons, have an s or $\bar{s}$ quark. The quark content of a proton is uud and the quark content of a neutron is ddu. Σ baryons are strange and have s or $\bar{s}$ quarks.

☐ Quarks are charged and so participate in electromagnetic interactions, but the charge on a quark is a fraction of the fundamental charge e, not a multiple of it. A u quark has a charge of $+2e/3$, a d quark has a charge of $-e/3$, and an s quark has a charge of $-e/3$. The charge on an antiquark is the negative of the charge on the associated quark. The charge on a hadron is simply the sum of the charges on its constituent quarks.

☐ Quark number is conserved in strong interactions. The strong interaction can create and destroy quark-antiquark pairs (such as $d\bar{d}$ or $u\bar{u}$) but it cannot change one type quark into another. The weak interaction, however, can change one type quark into another. In a beta decay of a proton $(p \rightarrow n + e^+ + \nu)$, a u quark is changed into a d quark. In a beta decay of a neutron $(n \rightarrow p + e^- + \bar{\nu})$, a d quark is changed into a u quark.

45–9 The Basic Forces and Messenger Particles

☐ In modern quantum theory, the basic interactions are pictured as exchanges of **messenger particles**. The messenger particles associated with the weak interaction are called W and Z, the messenger particle associated with the electromagnetic interaction is called the photon, and the messenger particles associated with the strong interaction are called **gluons**.

☐ When a particle emits a messenger particle, its rest energy is reduced. Nevertheless, it retains its character as long as it absorbs a messenger particle a short time later. Energy is not conserved during the interval between emission and absorption. For a given loss in energy ΔE, the time interval between emission and absorption must be less than the minimum predicted by the Heisenberg uncertainty principle. Thus, $\Delta t < h/\Delta E$. This limits the time for events, such as decays, and also limits the range of an interaction.

☐ Quarks interact with each other via the strong interaction, which involves the exchange of gluons. Inside a baryon or meson, the interaction force of two quarks is very weak but the interaction becomes extremely strong if the separation is larger than the particle size. Free quarks have never been observed. When a high-energy bombarding particle strikes a baryon or meson, other quarks are created (in quark-antiquark pairs) and these form combinations with existing quarks and are emitted as other mesons or baryons.

☐ The property of quarks that causes them to emit and absorb gluons is called **color**. It is somewhat similar to charge, the property that causes particles to absorb and emit photons and thus participate in the electromagnetic interaction. The chief difference is that a gluon carries color away from the quark that emits it and thereby changes the color of the quark. On the other hand, photons do not carry charge. The three types of color are called red, yellow, and blue.

☐ The condition that the net color of quarks in a hadron be neutral explains the limited quark combinations observed. A particle is color neutral only if it contains three quarks of different color, three antiquarks of different color, or a quark and antiquark of the same color.

☐ The basic interactions may be different manifestations of a single interaction. The electromagnetic and weak interactions have already been successfully unified into what is known as the **electroweak** interaction. A theory that unifies the strong and electroweak interactions is called a **grand unification theory** (GUT) and a theory that unifies all four interactions is called a **theory of everything** (TOE).

45–10 A Pause for Reflection

☐ All phenomena occurring at the present time can be explained in terms of the up quark, the down quark, the electron, the electron neutrino, and the messenger particles of their interactions. The other quarks and leptons are observed only in high energy experiments. However, they were present in the early universe and greatly influenced its evolution.

45–11 The Universe Is Expanding

☐ Distant astronomical objects are seen as they were when the light left them, perhaps more than 10^9 y ago.

☐ Most physicists believe that the universe started in a state of extremely high density and temperature, perhaps a highly concentrated mixture of quarks, leptons, and messenger particles. A "big bang" initiated an expansion (and cooling) that will continue for some time into the future, perhaps forever.

☐ Experimental evidence exists for the expansion of the universe and for the Big Bang. Doppler shift measurements are used to determine the speeds and directions of travel of distant galaxies. The galaxies are found to be moving away from each other. Hubble's law relates the speed with which a galaxy is receding from us to the distance between us and the galaxy. Mathematically it is $v = Hr$, where H is the Hubble parameter, about $80 \pm 17 \text{km}/(\text{s} \cdot \text{Mpc})$. Since a parsec is 3.256 ly, this is the same as $24.5 \text{ mm}/(\text{s} \cdot \text{ly})$. An observer in *any* part of the universe sees galaxies at the same distance receding with the same speed.

☐ If the rate of expansion of the universe has been constant, the reciprocal of the Hubble constant gives the age of the universe: about 1.2×10^{10} y.

45–12 The Cosmic Background Radiation

☐ The cosmic microwave background radiation fills the entire universe with an intensity that is nearly the same in every direction. There is no distinguishable source. It is believed to have originated shortly after the Big Bang. Although the background radiation started as a hot gas of highly energetic photons, in the gamma ray region of the electromagnetic spectrum, today it has a thermal spectrum corresponding to a temperature of about 2.7 K and is chiefly in the microwave region of the spectrum. The decrease in frequency accompanied the cooling of the universe.

45–13 Dark Matter

☐ Whether the universe will continue to expand indefinitely or will eventually collapse (perhaps in preparation for another big bang) depends on the amount of mass it contains. The universe seems to contain much more mass than is visible and the postulated invisible matter is called **dark matter**. The chief evidence for dark matter comes from a study of the rotation rates of distant galaxies.

45–14 The Big Bang

☐ The text describes seven eras in the history of the universe as it evolved from an extremely hot gas of quarks, leptons, and messenger particles to its present state.

NOTES:

NOTES

Notes